NUREG-0733

Analysis of Ultimate-Heat-Sink Spray Ponds

U.S. Nuclear Regulatory Commission

Office of Nuclear Reactor Regulation

R. Codell

NUREG-0733

Analysis of Ultimate-Heat-Sink Spray Ponds

Manuscript Completed: January 1981
Date Published: August 1981

R. Codell

**Division of Engineering
Office of Nuclear Reactor Regulation
U.S. Nuclear Regulatory Commission
Washington, D.C. 20555**

ABSTRACT

This report develops models which can be utilized in the design of certain types of spray ponds used in ultimate heat sinks at nuclear power plants, and ways in which the models may be employed to determine the design basis required by U.S. Nuclear Regulatory Commission Regulatory Guide 1.27.

The models of spray-pond performance are based on heat and mass transfer characteristics of drops in an environment whose humidity and velocity have been modified by the presence of the sprays. Drift loss from the sprays is estimated by a ballistics model.

The pond performance model is used first to scan a long-term weather record from a representative meteorological station in order to determine the periods of most adverse meteorology for cooling or evaporation. The identified periods are used in subsequent calculations to actually estimate the design-basis pond temperature. Additionally, methods are presented to correlate limited quantities of onsite data to the longer offsite record, and to estimate the recurrence interval of the design-basis meteorology chosen.

CONTENTS

CONTENTS (Continued)

CONTENTS (Continued)

CONTENTS (Continued)

FIGURES

FIGURES

TABLES

ACKNOWLEDGMENTS

The author of this report wishes to acknowledge several people who have made substantial contributions to its preparation. Much of the technical basis for the spray performance model can be traced directly to the work of Dennis Myers who also provided documentation and direct personal communication during development of the models.

Bill Nuttle (Massachusetts Institute of Technology) developed a large part of the meteorological scanning model for surface cooling ponds, which the author modified for spray ponds. He is also largely responsible for the drift-loss model.

I also wish to thank the Division of Technical Information and Document Control, for the careful editing of the text. Its editors suggested many useful changes. Beth Williams did an outstanding job typing the report, especially considering the number of difficult equations it contained.

Dennis Myers (Rodgers and Associates Corporation), Asem Elgawhary and Tim Morgan (Bechtel Power Corporation), and Bob Baird (Ford, Bacon, and Davis, Utah Corporation) reviewed the draft and made many good suggestions.

SYMBOLS

A = pond surface area, ft^2 or acres

A_c = total side area of the outermost segment, cm^2

A_d = cross-sectional area of the (assumed) spherical drop, cm

A_s = cross-sectional area of the spray field, cm^2

A_T = total area of top of spray field, cm^2

$A_{T,n}$ = top area of segment n, cm^2

$A_{T,N}$ = top area of segment N, cm^2

BDA = bone-dry air

C = cloud cover in tenths of the total sky obscured

C_d = drag coefficient for falling drops

C_p = heat capacity of water, cal/(gm °C)

C_{WA} = concentration of water in air in equilibrium at the temperature of the drop, gm water/cm^3 air

C_∞ = concentration of water in air in which the drop is immersed, gm water/cm^3 in air

    = molecular diffusivity of air, cm^2/sec

D = drop diameter, cm

D_i = mean diameter, cm

D_3 = "Sauter" mean diameter, cm

$D_{1/2}$ = mean drop diameter for spray performance calculations, cm

$drag$ = drag force, gm cm/sec^2

e_a = partial pressure of water vapor in the air, mm Hg

e_s = vapor pressure of water at the pond-surface temperature, mm Hg

E = equilibrium temperature, °F

ΔE = overall bias in pond temperature between the two data sets, °F

E_i = heat flowrate entering the segment, cal/sec

E_n = heat entering nth segment of sprays, cal/sec

f_i = fraction of drops in diameter range i whose diameter is D_i

F_b = buoyant force of rising air against the force of gravity, gm cm/sec^2

$F_{b,n}$ = buoyant force of rising air against force of gravity in segment n, gm cm/sec

$F(D)$ = probability density function for the drop-diameter distribution

$F_{d,n}$ = net drag force from falling droplets in segment n, gm cm/sec^2

$F(w)$ = wind function

g = acceleration of gravity, cm/sec^2

h_c = heat transfer coefficient for drop, cal/(sec cm^2 °C)

h_d = mass transfer coefficient for drop, cm/sec

h_n = heat flowrate of air leaving segment n, cal/sec

h_1 = heat flow rate of air leaving segment 1, cal/sec

H = humidity of air, gm water/cm^3 BDA

H_c = net rate of heat transfer from the pond due to conduction and convection, Btu/(ft^2 day)

H_n = humidity of air leaving segment n, gm water/gm BDA

H_0 = humidity of ambient air, gm water/gm BDA

$\dot{H}$ = rate of atmospheric heat transfer, Btu/(ft^2 day)

$\dot{H}_{AN}$ = net rate of longwave atmospheric radiation entering the pond, measured directly, Btu/(ft^2 day)

$\dot{H}_{BR}$ = net rate of back radiation leaving the pond surface, Btu/(ft^2 day)

$\dot{H}_C$ = net rate of heat flow from the pond due to conduction and convection, Btu/(ft^2 day)

$\dot{H}_E$ = net rate of heat loss due to evaporation, Btu/(ft^2 day)

$\dot{H}_{RJ}$ = net plant heat rejection, Btu/(ft^2 day)

$\dot{H}_{RJ,0}$ = steady-state heat load, Btu/(ft^2 day)

$\dot{H}_S$ = gross rate of solar radiation, Btu/(ft^2 day)

$\dot{H}_{SN}$ = net rate of shortwave solar radiation entering the pond, Btu/(ft^2 day)

H_{spray} = heat rejected by sprays, Btu/(ft^2 day)

I_i = evaporation from a single drop during its flight, gm

I_s = total daily solar radiation, Btu/(ft^2 day)

k = factor dependent on probability using Student's T distribution

k_a = thermal conductivity of air, cal/(cm sec °C)

K = equilibrium heat-transfer coefficient, Btu/(ft^2 hr °F)

m = mass of drop, gm

M = sample mean

$M_{i,n}$ = mass flowrate of water vapor entering segment n from the spray, from drops of diameter range i, gm/sec

M_n = total mass flowrate of water vapor entering segment n, gm/sec

M_w = molecular weight of water, 18 gm/gm mole

M_y = momentum of drop in vertical direction, gm cm/sec

$M_{y,i}$ = net downward momentum of the falling drops of diameter range i, gm cm/sec

p = atmospheric pressure, mm Hg

p_w = vapor pressure of water, mm Hg

P = probability

P_i = plotting position for ranked annual maximum values in probability coordinates

Pr = Prandtl number

q = evaporation rate, Btu/hr

$Q_{w,n}$ = flowrate of water into the nth section

Q = flowrate of water to spray field, cm^3/sec or ft^3/hr

ΔQ = evaporation correction factor, ft^3/hr

Q_A = flowrate of BDA, gm BDA/sec

$Q_{A,n}$ = flowrate of BDA leaving segment n, gm BDA/sec

$Q_{A,N}$ = net outward flowrate of BDA leaving the innermost segment N of the spray field, gm BDA/sec

$Q_{A,0}$ = quantity of BDA entering the first segment of the spray field, gm BDA/sec

$Q_{T,n}$ = quantity of BDA leaving top of segment n in LWS model, gm BDA/sec

r = drop radius, cm

r^2 = coefficient of determination

r_i = particular average radius of drop, cm

R_c = cooling range of the sprays

Re = Reynold's number of drop

R_g = universal gas constant, 82.02 cm^3 atm/(gm mole $^\circ$K)

S = standard deviation

Sc = Schmidt number

t = time, sec or hr

t_f = time for drop to fall to water surface, sec

t_1 = one-half the length of daylight per day, hr

t_0 = time of the observation (in hours before or after midday)

T = temperature of drop, $^\circ$C or $^\circ$K

ΔT	= correction factor for peak temperature
T_A	= air temperature, °F or °C
$T_{A,0}$	= ambient air temperature, °C
T_{HOT}	= temperature of the drop when it left the nozzle, °C or °F
T_{max}	= highest observed value of pond water temperature, °F
T_{100}	= 100-yr recurrence interval pond temperature, °F
T_s	= pond surface temperature, °F
T'_s	= pond ambient temperature, °F
ΔT_v	= "virtual" temperature difference between the pond surface water and air above the pond, °F
$T_{w,n}$	= temperature of liquid water leaving segment n, °C
T_W	= wet-bulb temperature, °F
$T_{w,1}$	= temperature of liquid water leaving segment 1, °F
u	= velocity of drop in x direction, cm/sec
u'	= ambient air velocity component, cm/sec
v	= velocity of drop in y direction, cm/sec
v'	= ambient air velocity component, cm/sec
v'_n	= net upward- or downward-induced air velocity, cm/sec
V	= absolute velocity of drop relative to air, cm/sec
V_h	= humid volume of the ambient air, cm³/gm BDA
$V_{h,N}$	= humid volume of the air in segment N, cm³/gm BDA
V_i	= volume of drop in size range i, cm³
V_p	= pond volume, ft³
w	= windspeed perpendicular to the pond, either naturally impinging or induced, cm/sec
w_0	= induced windspeed at the circumference, cm/sec
W	= flowrate through pond or sprays, ft³/hr
W_b	= flowrate of the blowdown or leakage stream, ft³/hr
W_{drift}	= water loss attributable to drift, cm³/sec
W_e	= evaporation rate, ft³/hr
W_l	= total water loss attributable to sprays, cm³/sec
W_{max}	= maximum observed value of evaporation, ft³/30 days
W_{100}	= 100-yr recurrence interval 30-day evaporation, ft³/30 days
W_{spray}	= rate of water evaporated from all drops in the spray field, ft³/hr

ΔZ = one-half the height of the spray field, cm

α = convergence parameter

η = spray efficiency

θ = excess temperature, °F

λ = heat of vaporization for water, cal/gm

μ = viscosity of air, gm/(cm sec)

ρ = density of water, lb/ft^3 or gm/cm^3

ρ_A = density of air , gm/cm^3

$\overline{\Delta\rho}_A$ = average density difference between the air in segment n and the ambient air, gm/cm^3

$\overline{\rho}_{A,n}$ = average density of the air in segment n, gm/cm^3

σ = standard error

ANALYSIS OF ULTIMATE-HEAT-SINK SPRAY PONDS

1 INTRODUCTION

The ultimate heat sink (UHS) is defined as the complex of cooling-water sources necessary to safely shut down and cool down a nuclear power plant. Cooling ponds, spray ponds, and mechanical draft cooling towers are some examples of the types of ultimate heat sinks in use today.

The U.S. Nuclear Regulatory Commission (NRC) has set forth in Regulatory Guide 1.27 (Ref. 1) the following positions on the design of ultimate heat sinks: (1) The ultimate heat sink must be able to dissipate the heat of a design-basis accident (for example, loss-of-coolant accident) of one unit plus the heat of a safe shutdown and cooldown of all other units it serves. (2) The heat sink must provide a 30-day supply of cooling water at or below the design-basis temperature for all safety-related equipment. (3) The system must be shown to be capable of performing under the meteorologic conditions leading to the worst cooling performance and the conditions leading to the highest water loss.

This report identifies methods that may be used to select the most severe combinations of controlling meteorological parameters for a spray-cooling pond of conventional design. The procedure scans a long-term weather record, which is usually available from the National Weather Service for a nearby station, and predicts the period for which either pond temperature or water loss would be maximized for a hydraulically simple spray pond. The principle of linear superposition is used to develop a procedure that allows the peak ambient pond temperature to be superimposed on the peak "excess" temperature, due to plant-heat rejection. This procedure determines the timing within the weather record of the peak ambient pond temperature. The true peak can then be determined in a subsequent, more-rigorous calculation.

Maximum 30-day water loss is determined directly from the scanning model.

The data-scanning procedure requires a data record on the order of tens of years to be effective. Since these data will usually come from somewhere other than the site itself (such as a nearby airport), methods to compare these data with the limited onsite data are developed so that the adequacy or at least the conservatism of the offsite data can be established. Conservative correction factors to be added to the final results are suggested.

These models and methods, provided as useful tools for UHS analyses of spray cooling ponds, are intended as guidelines only. Use of these methods does not automatically assure NRC approval, nor are they required procedures for nuclear-power-plant licensing. Furthermore NRC does not, by publishing this guidance, wish to discourage independent assessments of UHS performance or furtherance of the state of the art.

2 SPRAY-POND HEAT AND MASS TRANSFER PERFORMANCE MODELS

A set of models which consider the interaction of sprayed water with air in a spray pond has been developed to calculate cooling and water-loss performance. The models are developed along the line of other models of spray-pond performance (Refs. 2 and 3 and D. M. Myers, personal communication, 1976) and have been tested with field data on prototype ponds. These models form the bases of the analytical methods of spray-pond analysis.

The performance model is developed in two parts:

(1) A "microscale" submodel which considers the heat, mass, and momentum transfer of a single drop as it falls through the surrounding air.

(2) "Macroscale" submodels which consider the modification of the surrounding air resulting from the heat, mass, and momentum transfer from many drops in different parts of the spray field.

The microscale and macroscale submodels are combined into a model of performance of the entire spray field. This spray-field model may then be combined with a submodel of the pond itself to simulate the performance of the total UHS system.

2.1 Microscale Submodel

This portion of the model considers the heat, mass, and momentum transfer from a single water drop with the surrounding air.

2.1.1 Drop Motion

The motion of the drop after it leaves the spray nozzle is approximated by the classic ballistic problem as described in Figure 2.1. Drops leave the nozzle

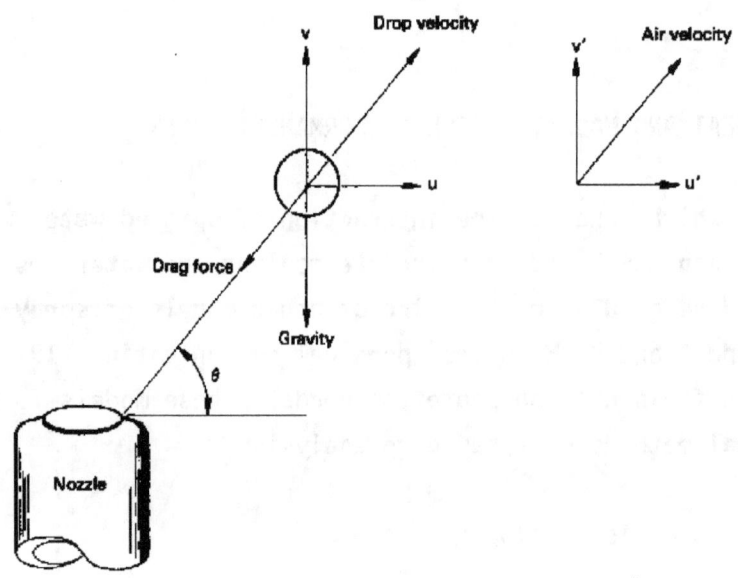

Figure 2.1 Ballistics of a drop leaving a spray nozzle

at an angle θ to the horizontal. After leaving the nozzle, the drop is subjected to the force of gravity and drag from the air. The motion of the drop is represented by the following differential equations:

$$\frac{du}{dt} = -\frac{C_d A_d \rho_A (u - u')V}{m} \tag{2.1}$$

$$\frac{dv}{dt} = -\frac{C_d A_d \rho_A (v - v')V}{m} - g \tag{2.2}$$

where

u = velocity of drop in x direction, cm/sec

v = velocity of drop in y direction, cm/sec

t = time, sec

C_d = drag coefficient for falling drops

A_d = cross-sectional area of drop, cm^2

ρ_A = air density, gm/cm^3

u', v' = ambient air velocity components, cm/sec

V = absolute velocity of drop relative to air

m = mass of drop, gm

g = acceleration of gravity, cm/sec^2

C_d, a drag coefficient for falling drops, is a function of Reynold's number Re. An approximation of C_d as a function of Re for rigid spheres is suggested by Bird, Stewart, and Lightfoot (Ref. 4):

For Re < 2

$$C_d = \frac{24}{Re} \qquad (2.3)$$

For 2 < Re < 500

$$C_d = \frac{18.5}{Re^{0.6}} \qquad (2.4)$$

For Re > 500

$$C_d = 0.44 \qquad (2.5)$$

Reynold's number is defined in the following relationship:

$$Re = \frac{2rV\rho}{\mu}$$

where

r = drop radius, cm

μ = viscosity of air, gm/(cm sec)

and V and ρ are as previously defined.

2.1.2 Heat and Mass Transfer Relations

The falling drop exchanges heat and mass with the surrounding air. The rate of change of the drop's temperature may be expressed in terms of the following differential equation (Ref. 2):

$$\frac{dT}{dt} = - \frac{1}{\frac{4}{3}C_p\rho\pi r^3} \left[4\pi r^2 h_d \left(C_{WA} - C_\infty\right)\lambda + 4\pi r^2 h_c \left(T - T_{A,\infty}\right) \right] \qquad (2.6)$$

where

T = temperature of the drop, °C

C_p = heat capacity of water, cal/(gm °C)

ρ = density of water, gm/cm^3

h_d = mass transfer coefficient, cm/sec

C_{WA} = concentration of water in air in equilibrium at the temperature of the drop, gm water/cm^3 air

C_∞ = concentration of water in air in which the drop is immersed, gm water/cm^3 air

λ = heat of vaporization of water, cal/gm

h_c = heat-transfer coefficient, cal/(sec cm^2 °C)

$T_{A,\infty}$ = temperature of the air in which the drop is immersed, °C

and t and r are as previously defined.

The heat and mass transfer coefficients h_c and h_d, respectively, are based on the classic work of Ranz and Marshall (Ref. 5) on pendant drops. The heat-transfer coefficient h_c has been empirically determined to be:

$$h_c = \frac{k_a}{r}(1 + 0.3Pr^{1/3}Re^{1/2}) \text{ cal/(sec cm}^2 \text{ °C)} \qquad (2.7)$$

where

k_a = thermal conductivity of air, cal/(sec cm °C)

Pr = Prandtl number

Re = Reynolds number

and h_c and r are as previously defined.

Similarly, the mass transfer coefficient has been empirically determined to be:

$$h_d = \frac{\mathcal{D}}{r}(1 + 0.3Sc^{1/3}Re^{1/2}) \text{ cm/sec} \qquad (2.8)$$

where

$\mathcal{D}$ = molecular diffusivity of air, cm^2/sec

Sc = Schmidt number

and h_d, r, and Re are as previously defined.

The concentration C_∞ is determined from the ideal gas law:

$$C_\infty = \frac{p_w M_w}{R_g T} \qquad (2.9)$$

where

p_w = vapor pressure of water, atm

M_w = molecular weight of water, 18 gm/gm mole

R_g = universal gas constant, 82.02 cm^3 atm/(gm mole °K)

T = absolute temperature of the drop, °K

and C_∞ is as previously defined.

The parameters ρ, μ, Pr, Sc, D and k_a (all previously defined) are thermodynamic properties of the air-water system. For the present purposes, these have been expressed by the following empirical relationships in terms of the absolute temperature of air, T_A, °K (Refs. 2 and 6):

$$\mu = 2.7936 \times 10^{-6} T_A^{0.73617} \text{ gm/(cm sec)} \qquad (2.10)$$

$$\rho = 0.353 T_A^{-1} \text{ gm/cm}^3 \qquad (2.11)$$

$$Pr = 0.93176 \ T_A{}^{-0.042784} \qquad (2.12)$$

$$Sc = 2.2705 \ T_A{}^{-0.21398} \qquad (2.13)$$

$$\mathcal{D} = 5.8758 \times 10^{-6} \ T_A{}^{1.8615} \ cm^2/sec \qquad (2.14)$$

$$k_a = 3.9273 \times 10^{-7} \ T_A{}^{0.88315} \ cal/(cm \ sec \ {}^{\circ}C) \qquad (2.15)$$

The vapor pressure of water may be expressed in terms of the absolute water temperature of the drop, T $({}^{\circ}K)$:

$$\ln p_w = (71.02499 - 7381.6477/T - 9.0993037 \ \ln T$$
$$+ \ 0.0070831558 \ T) \ atm \qquad (2.16)$$

2.1.3 Momentum Transfer

The falling water drops will impart momentum to the surrounding air because of drag. Since the spray from a single nozzle will be axially symmetrical, the net momentum in the x direction should be approximately zero.* In typical UHS designs the net momentum change in the vertical direction due to the drag from the drops will be in the downward direction. The net momentum is defined by the integral:

$$M_y = \int_0^{t_f} \frac{drag}{V} \ dt \quad gm \ cm/sec \qquad (2.17)$$

where

t_f = time for drop to fall to water surface

drag = drag force, $(gm \ cm)/sec^2$

---

*In this analysis, oriented spray nozzles which are purposely arranged to induce a lateral flow are not considered.

2.1.4 Solution of Microscale Equations

The above equations are solved simultaneously with numerical integration in a
fourth-order Runge-Kutta scheme. Mass, heat, and momentum transfer are calculated for a single drop, specifying as inputs the drop radius, the initial
velocity from the nozzle, the spray angle, the height of the nozzle above the
water surface, the sprayed temperature, and the temperature and humidity of
the surrounding air. The outputs from this submodel are subsequently used in
the macroscale submodel.

2.2 Macroscale Submodels

The performance of a single isolated spray nozzle might be adequately predicted
by the microscale model alone. When many spray nozzles are arranged into a
spray field, however, consideration must be given to the modification of the
atmospheric environment in which the nozzle is immersed because of neighboring
spray nozzles. The temperature and humidity of the air in the interior of a
spray field are both raised and will lead to diminished spray performance with
respect to an isolated nozzle in unaffected air. In addition, heated, humidified air is less dense than cooler, drier air. Therefore, it is likely that
complicated convection currents will be generated, which may also be affected
by the drag forces of the falling drops.

There are separate macroscale models dealing with high- and low-windspeed
conditions. The high-speed model assumes that the momentum exchange in the
pond due to drag and buoyancy are much less important than that due to the wind
blowing through the spray field. The low-speed model assumes that the opposite
is the case. The transfer of the air through the spray field is self-induced.

Both models are run at the same time in the simulation, since for some cases
of high-heat loadings, natural convection might be greater than wind-induced
convection. The higher performance model is then chosen as being representative
of the spray field for that time interval.

2.2.1 High-Windspeed Submodel

The spray field is represented by a rectangular volume, in which the density of sprayed drops is great, as represented in Figure 2.2. The rectangular volume is divided into N equal segments. Each segment is then considered to be a compartment whose air temperature and humidity are determined by the preceding segment, as depicted in Figure 2.3.

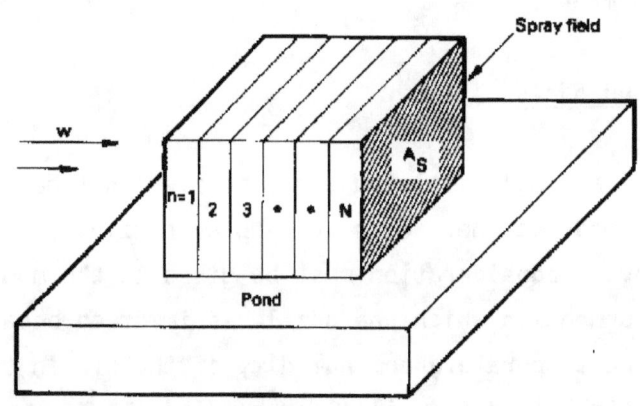

Figure 2.2 Segmentation of spray field for high-windspeed model

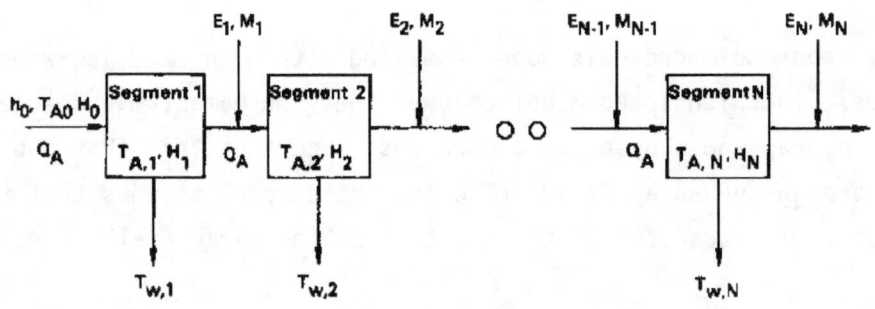

Figure 2.3 Compartment model of spray field for high windspeed

Ambient air of humidity H_0 (gm water/gm dry air) and temperature $T_{A,0}$ (°C) enters the first segment of the spray field at a volumetric rate wA_s cm³/sec, where w is the windspeed perpendicular to the long axis of the pond (cm/sec) and A_s is the cross-sectional area of the spray field (cm^2). It is convenient to perform all mass and heat-transfer calculations on a "bone dry air" (BDA) basis (Ref. 6). The "humid volume" V_h is defined as the volume occupied by a parcel of air whose dry weight is 1 gm and at a pressure of 1 atm:

$$V_h = \left(81.86T_{A,0} + 22,387\right)\left(\frac{1}{29} + \frac{H_0}{18}\right) \text{ cm}^3/\text{gm BDA} \qquad (2.18)$$

The quantity of BDA, Q_A, entering and passing through every segment of the pond (flow rate) is, therefore:

$$Q_A = w \times \frac{A_s}{\left(81.86T_{A,0} + 22,387\right)}\bigg/\left(\frac{1}{29} + \frac{H_0}{18}\right) \text{ gm BDA} \qquad (2.19)$$

The concentration of water in air C_{WA} anywhere in the pond is related to the humidity H and temperature T_A by the relationship:

$$C_{WA} = \frac{H}{\left(81.86T_A + 22,387\right)\left(\frac{1}{29} + \frac{H}{18}\right)} \text{ gm water/cm}^3 \text{ wet air} \qquad (2.20)$$

For a particular segment n, it can be assumed that the humidity and air temperature are determined only by what left the segment upwind, providing that all other parameters of the system, such as initial drop velocity, spray angle, nozzle height, and hot-water temperature, are known. Subroutine SPRAY is then called several times for each segment n or to solve the microscale equations of heat and mass transfer from drops over a range of radii whose distribution is typical of the particular nozzle design employed.

For drops of a particular average radius r_i (cm), the heat entering the segment, E_i, is proportional to the fraction of drops in that diameter range (diameter D_i); f_i, the flowrate of water into the nth section q_{wn}; and the difference between the temperature of the drop when it left the nozzle, T_{HOT}, and when it reached the pond surface, $T_i(^\circ C)$:

$$E_i = \rho C_p q_{wn} f_i \frac{T_{HOT} - T_i}{\frac{4}{3}\pi r_i^3} \text{ cal/sec} \tag{2.21}$$

The total rate of heat entering pond segment n is therefore:

$$E_n = \sum E_i = \frac{\rho C_p Q_{wn}}{\frac{4}{3}\pi} \sum_{i=1}^{j} \frac{T_{HOT} - T_i}{f_i r_i^3} \text{ cal/sec} \tag{2.22}$$

where j is the number of drop-diameter ranges used.

The heat flowrate in the air leaving segment n (and entering segment n + 1) is therefore:

$$h_{n+1} = h_n + E_n \text{ cal/sec} \tag{2.23}$$

where h_n = heat flow rate leaving segment n, cal/sec.

Liquid water leaving the segment is of temperature:

$$T_{w,n} = \sum_{i=1}^{j} f_i T_i \,^\circ C \tag{2.24}$$

For drops of a particular average radius r_i (cm), the mass flowrate entering the segment from the sprays will be:

$$M_{i,n} = \frac{f_i q_{wn} I_i}{\frac{4}{3}\pi r_i^3} \text{ gm/sec} \tag{2.25}$$

where I_i is the evaporation from a single drop during its flight, in grams.

The total mass flowrate of water vapor entering segment n is therefore:

$$M_n = \sum_{i=1}^{j} M_{i,n} = \frac{3q_{wn}}{\frac{4}{3}\pi} \sum_{i=1}^{j} \frac{f_i I_i}{r_i^3} \text{ gm/sec} \qquad (2.26)$$

Adding M_n gm/sec of water vapor to the air leaving the segment n increases the humidity of segment n + 1 by the following amount:

$$H_{n+1} = H_n + \frac{M_n}{Q_A} \text{ gm water/gm BDA} \qquad (2.27)$$

The temperature of the air leaving one segment and entering the next reflects the added heat and moisture:

$$T_{A,n+1} = \frac{\frac{h_{n+1}}{Q_A} - H_{n+1}\lambda}{0.24 + 0.45\,H_{n+1}} \quad °C \qquad (2.28)$$

Calculations continue with segment n + 1, and step through all pond segments.

The properties of the air in the first segment are determined by the ambient air temperature $T_{A,0}$ and humidity H_0:

$$T_{A,1} = T_{A,0} \;°C$$
$$H_1 = H_0 \text{ gm water/gm BDA}$$
$$h_1 = Q_A\left[0.24 T_{A,0} + H_0\left(\lambda \pm 0.45 T_{A,0}\right)\right] \text{ cal/sec} \qquad (2.29)$$

The total cooling performance of the spray field is simply the average cooling from all sections:

$$\text{Range} = \frac{\displaystyle\sum_{n=1}^{N} q_{wn}\left(T_{HOT} - T_{w,n}\right)}{\displaystyle\sum_{n=1}^{N} q_{w,n}} \quad °C \qquad (2.30)$$

Cooling performance may also be expressed in terms of "efficiency" of approach to wet-bulb temperature:

$$\eta = \frac{\text{range}}{\left(T_{HOT} - T_W\right)} \times 100 \quad \text{percent} \qquad (2.31)$$

2.2.2 Low-Windspeed Macroscale Submodel

At low ambient windspeeds, the flow of air through the spray field is largely controlled by two mechanisms: drag from the spray droplets and buoyancy of the heated, humidified air. Since the spray-field arrangements in most conventional spray fields are already evenly distributed and symmetrical, it would appear that there would be little net effect of the spray droplet drag in the lateral direction. There would be a net downward drag due to the falling drops.*

In a conventional spray pond under loads typical of UHS service, buoyancy is the dominant force in the low-windspeed case.

For the low-windspeed model, the spray field is sectioned into N rectangular cylinders of equal volume as shown in Figure 2.4 (Ref. 3 and D. M. Myers, personal communication, 1976). Air enters the segment from all four sides,

---

*However, at least one spray-equipment manufacturer, Ecolaire (Ref. 7), is marketing an oriented spray-field arrangement which induces the circulation of air laterally.

and leaves the segment to enter the next segment after being heated and
humidified by the sprays. Unlike the high-windspeed model, however, air also
leaves through the top of the segment because of buoyancy. Each segment is then
considered to be a compartment whose air temperature, humidity, and air-flow
rate are determined by the heat and mass transfer of the segment itself and
the previous and next segments as depicted in Figure 2.5.

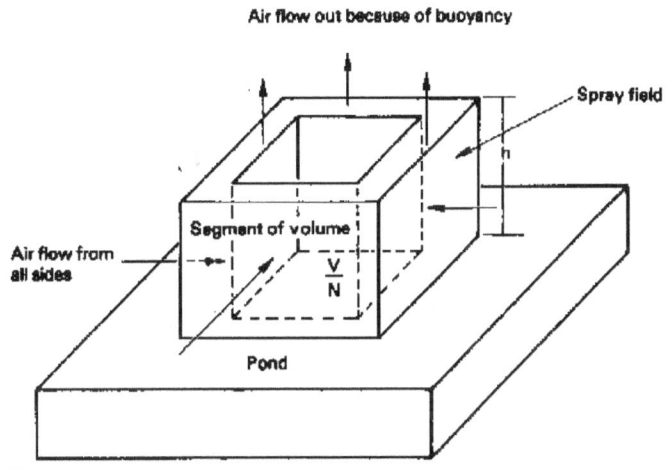

**Figure 2.4   Segmentation of spray field for
low-windspeed model**

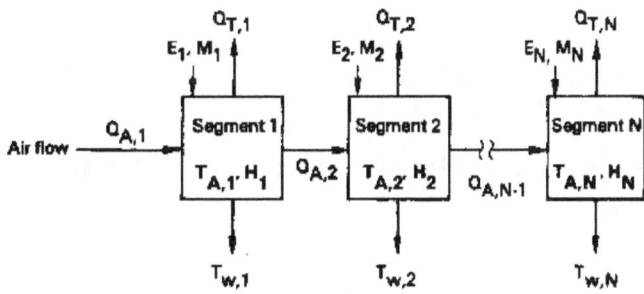

**Figure 2.5   Compartment model of spray field for
low windspeed**

2.2.2.1 Material and Energy Balances of Segment n

If a control volume is drawn around segment n, the relationships between the air
and water streams can be defined. The flow of air is described on a BDA basis:

$$Q_{A,n} = \text{air leaving segment } n = Q_{A,n+1} + Q_{T,n} \text{ gm BDA/sec} \tag{2.32}$$

and the water vapor entering segment n will be:

$$M_n = \frac{3q_{wn}}{4\pi} \sum_{i=1}^{j} \frac{f_i I_i}{r_i^3} \text{ gm/sec} \tag{2.33}$$

Adding M_n gm/sec of water vapor to the air leaving segment n increases the humidity of segment $n + 1$:

$$H_{n+1} = H_n + \frac{M_n}{Q_{A,n}} \tag{2.34}$$

The temperature of the air leaving segment n and entering the next is modified by the added heat and moisture:

$$T_{n+1} = \frac{\dfrac{h_{n+1}}{Q_{A,n}} - H_{n+1} \lambda}{0.24 + 0.45 \, H_{n+1}} \quad ^\circ C \tag{2.35}$$

where λ = heat of vaporization, cal/gm.

The quantity of BDA entering the first segment Q_{A_0} of the spray field is defined to be:

$$Q_{A_0} = \frac{w_0 \times A_c}{\left(81.86 T_{A,0} + 22,387\right)\left(\dfrac{1}{29} + \dfrac{H_0}{18}\right)} \text{ gm BDA/sec} \tag{2.36}$$

where

w_0 = induced windspeed at the circumference, cm/sec

A_c = total side area of the outermost segment, cm^2

and $T_{A,0}$ and H_0 are as previously defined.

Air leaving the last segment can leave only through the top, so:

$$Q_{A,N} = 0 \qquad\qquad (2.37)$$

2.2.2.2 Momentum Balance

The movement of air and water vapor through the spray field is controlled by complicated aerodynamic effects. In the grossest sense, however, a balance of vertical momentum, i.e., Bernoulli's equation (Ref. 4), can be used to represent the movement of air streams. For any segment n, the vertical momentum of the entering and leaving streams of air is defined by the following equations:

(1) Force of air leaving top of segment:

$$v_n'^2 \bar{\rho}_{A,n} A_{T,n} \qquad\qquad (2.38)$$

where

v_n' = upward velocity of the air in segment n, cm/sec

$\bar{\rho}_{A,n}$ = average density of the air in segment n, gm/cm^3

$A_{T,n}$ = top area of segment n, cm^2

(2) The buoyant force of rising air against the force of gravity in segment n:

$$F_{b,n} = A_{T,n} g \overline{\Delta\rho}_{A,n} \Delta Z \text{ gm cm/sec}^2 \qquad\qquad (2.39)$$

where

$\overline{\Delta\rho}_{A,n}$ = average density difference between the air in segment n and the ambient air, gm/cm^3

ΔZ = one-half the height of the spray field, cm

and $A_{T,n}$, g, and $\bar{\rho}_{A,n}$ are as previously defined.

2.2.2.3 Net Drag Force From Falling Droplets in Segment n

$$F_{d,n} = \sum_i \frac{f_i M_{y,i} Q}{V_i A_T} \text{ gm cm/sec}^2 \qquad (2.40)$$

where

$M_{y,i}$ = net downward momentum of each of the falling drops in diameter range i (from Eq. 2.17), gm cm/sec

Q = flowrate of water to spray field, cm³/sec

V_i = volume of drop in size range i, cm³

A_T = total top surface area of the spray field, cm²

The net upward or downward air velocity of the air in segment n, v_n' (cm/sec) is found by solving one of the following two expressions:

$$v_n' = \sqrt{(F_{b,n} + F_{d,n})/\rho_A} \text{ if } (F_{b,n} + F_{d,n}) > 0 \qquad (2.41)$$

for upward velocity or

$$v_n' = -\sqrt{-(F_{b,n} + F_{d,n})/\rho_A} \text{ if } (F_{b,n} + F_{d,n}) < 0 \qquad (2.42)$$

for downward velocity.

2.2.2.3 Solving for Air Flow

The velocity of air leaving each segment is calculated at each iteration based on the temperature and humidity of the segments in the previous iteration. The calculation of mass transport through the spray field starts at the inner-most segment. The net outward flowrate of BDA leaving the innermost segment N of the spray field is:

$$Q_{A,N} = \frac{v_N' A_{T,N}}{V_{h,N}} \text{ if } v_N' \text{ is positive} \qquad (2.43)$$

$$Q_{A,N} = \frac{v_N' A_{T,N}}{V_h} \text{ if } v_N' \text{ is negative} \tag{2.44}$$

where

$A_{T,N}$ = top area of segment N, cm^2

$V_{h,N}$ = humid volume of the air within segment N, cm^3/gm BDA

V_h = humid volume of the ambient air, cm^3/gm BDA

and v_N' is as previously defined.

The flowrate of BDA for all other segments $Q_{A,n}$ is calculated by stepping from the innermost segment outward:

$$Q_{A,n} = Q_{A,n+1} + \frac{v_n' A_{T,n}}{V_{h,n}} \text{ if } v_n' \text{ is positive} \tag{2.45}$$

$$Q_{A,n} = Q_{A,n+1} + \frac{v_n' A_{T,n}}{V_h} \text{ if } v_n' \text{ is negative} \tag{2.46}$$

The temperature and humidity in each segment are next recomputed based on the new estimate of flowrate of BDA starting with the outermost segment and working in. The enthalpy of air entering the first segment is simply that of the ambient air H_0.

2.2.2.4 Convergence of Iterative Solution

The computations for the LWS (low-windspeed) model outlined above are iterative. The flowrate of air and water vapor depends on the computed temperature and humidity in each segment. Conversely, the temperature and humidity depend on the flow of air through the spray field. The computations proceed iteratively until the differences of temperature, humidity, and air flow between two computations are smaller than a certain tolerance.

Under certain circumstances, convergence may be very difficult. For example, a poor initiation of the computation may cause the first calculated flowrates to be very small, which in turn would cause the subsequently calculated temperatures to be very large. Because the equations are highly nonlinear, a wide initial oscillation may drive the iterative calculations beyond the region of convergence and into a region of divergence where the solution will degenerate or "blow up."

Other factors contribute to the divergence of the solution of the LWS model. The effect of the downward drag of falling drops seems to destabilize the calculation, especially if the net flow from any segment were to be downward instead of upward.

2.2.2.5 Measures To Aid Convergence

It is possible to assure convergence of the LWS model in almost every case by imposing several computational restrictions:

(1) Allow only positive (upward) air flow from each segment.
(2) Eliminate vertical drag as a force in the momentum balance.
(3) Introduce "damping" to smooth out oscillations.

Steps 1 and 2 above are compromises which could affect the computation accuracy. The effect of these restrictions on the resultant performances is shown later to be minor and in fact appears to improve the model's comparison to field data.

Damping is a computational trick which has the effect of smoothing large oscillations, but whose influence disappears at steady state (Ref. 8). The temperature and humidity in the ith segment are damped in the following manner:

$$T_{A,i}^{k'} = T_{A,i}^{k} - \alpha \left(T_{A,i}^{k-2} - 2T_{A,i}^{k-1} + T_{A,i}^{k} \right) \tag{2.47}$$

$$H_{i}^{k'} = H_{i}^{k} - \alpha \left(H_{i}^{k-2} - 2H_{i}^{k-1} + H_{i}^{k} \right) \tag{2.48}$$

where

$T_{A,i}^{k'}$ = smoothed value of air temperature in the segment $T_{A,i}^{k}$

2-18

$H_i^{k'}$ = smoothed value of the humidity in the segment H_i^k

α = convergence coefficient. Typically, its value should be about 0.05 to 0.1, but other values may be used.

The superscript k represents the present iteration; the superscripts k - 1 and k - 2 represent the previous two iterations, respectively.

An example of the effect of damping is illustrated in Figure 2.6. The temperature of the innermost segment oscillates around the steady-state value, but appears to converge faster with damping. The results for the damping case are identical until the third iteration since the damping factor depends on having results from two previous iterations.

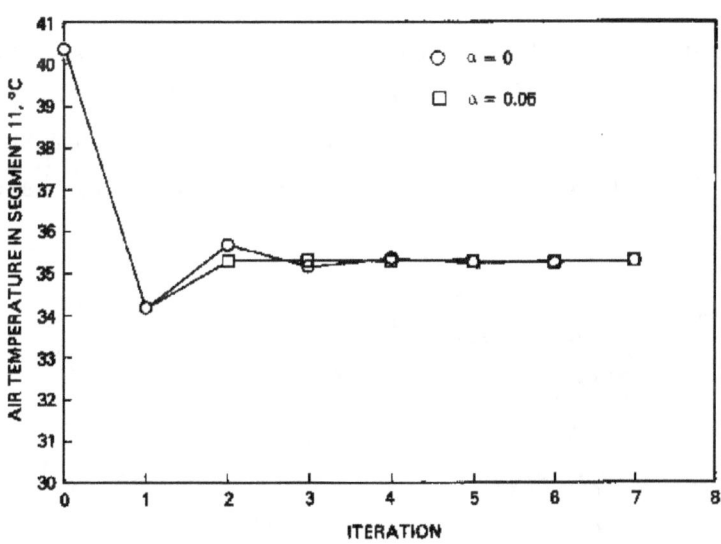

Figure 2.6 Convergence of low-windspeed model with and without damping

2.3 Comparison With Field Data

The results of the spray performance models were compared with available data on spray-pond performance.* Two sets of data were generally available at the

---

*Actually, the "mean drop diameter" simplification was taken, as developed in Section 3, but the results are shown to be nearly identical.

2-19

time the comparison was made: (a) the Canadys test data (Ref. 9), and (b) the Rancho Seco spray-pond confirmatory tests (Ref. 10). Both data sets considered only the instantaneous cooling of the sprays, and did not attempt to include other heat-transfer mechanisms, such as cooling from the pond surface. The Canadys data were gathered on an operating spray-cooling pond used for condensor cooling at a fossil-fuel electric station in South Carolina. The Rancho Seco data were gathered at an actual UHS spray pond in California during a preoperational test requested by the NRC. The Rancho Seco tests were designed specifically to determine the performance of the spray field, while the Candys tests considered the performance of the pond as a whole, including heat transfer from the pond's surface. The Rancho Seco data are more appropriate for the present comparison.

2.3.1 Canadys Data Comparison

The Canadys spray pond is shown in Figure 2.7. Not all of the information on the basic physical parameters of the Canadys spray pond could be found, and some parameters had to be inferred. For example, the height of the nozzles, the height of the sprayed water, the nozzle distribution, and the drop-diameter

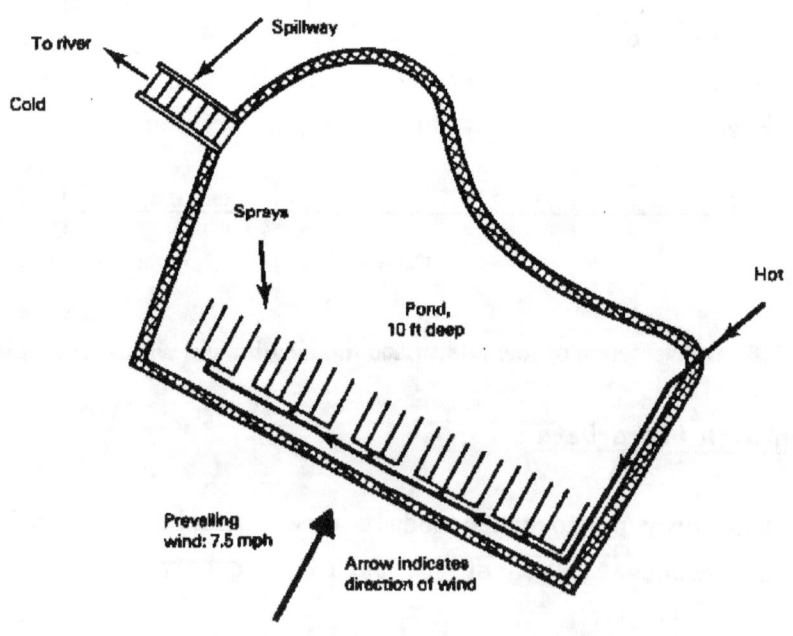

Figure 2.7   Canadys spray-cooling pond

distribution of the sprays (in 10 divisions shown in Table 2.5) were taken as those for the Spraco 1751 nozzle and recommended layout, although the design probably was somewhat different (Ref. 11). It should be noted that the performance models can be used with any nozzle as long as the drop-diameter distribution is known. Necessary pond parameters are shown in Table 2.1. Table 2.2 contains the measured atmospheric variables and pond performance in terms of spray efficiency η, as well as the predicted performance from the HWS model. Figure 2.8 plots the predicted efficiency versus the measured efficiency. There is a great deal of scatter evident from Figure 2.8, but the points distributed on the diagonal, indicating no systematic bias.

Table 2.1   Physical Characteristics of Canadys
Spray Pond Used in Spray-Field Model

Variable	Measurement
Length of spray field	304.8 m
Width of spray field	30.48 m
Height of spray field	3.66 m
Initial drop velocity	6.67 m/sec
Angle of drop with respect to horizon	76°
Height of nozzles from water surface	1.52 m
Barometric pressure	29.92 in. Hg
Flowrate through all nozzles	11,400 liters/sec

It should be noted that the Canadys pond has a sprayed-water loading about twice that recommended by spray-nozzle manufacturers. The cooling efficiency of this pond and that predicted by the NRC model were well below the efficiencies predicted by conventional techniques before the pond was constructed.

2.3.2   Rancho Seco Data

The Rancho Seco pond is shown in Figure 2.9. This pond incorporates a standard Spraco design for spray configuration and the employment of the 1751 nozzle. Most operational characteristics of the pond were well documented. The basic

Table 2.2 Measured Atmospheric Parameters and Spray Efficiency, and Efficiency Predicted From High-Windspeed Model With Drag Terms Included

$T_w - °C$	$T_A, °C$	$T_{HOT}, °C$	w, cm/sec	$\eta_{measured}$	$\eta_{predicted}$
25.4	30.4	43.6	163.2	0.443	0.250
27.1	35.6	45.0	244.8	0.248	0.334
27.1	34.6	44.7	204.0	0.279	0.301
23.2	24.7	44.2	244.8	0.275	0.310
25.8	27.2	41.7	201.0	0.346	0.279
26.1	30.3	42.8	191.0	0.270	0.276
26.1	31.7	43.6	201.0	0.325	0.288
24.2	27.5	43.3	163.2	0.257	0.244
26.6	31.3	42.2	175.9	0.320	0.261
25.4	28.5	44.2	163.2	0.252	0.253
26.8	31.1	43.6	226.1	0.265	0.312
25.6	35.2	45.3	163.3	0.198	0.257
26.6	30.9	45.6	276.4	0.263	0.353
27.4	34.1	44.4	271.0	0.351	0.350
25.4	36.7	45.6	246.4	0.252	0.328
26.5	36.1	44.4	246.4	0.302	0.329
21.3	25.8	43.9	427.2	0.339	0.378
22.1	25.0	44.4	305.1	0.372	0.339
21.6	24.3	43.9	276.4	0.343	0.319
20.8	24.4	44.4	226.1	0.335	0.288
16.8	21.7	36.1	376.9	0.346	0.305
17.8	24.7	37.8	414.6	0.275	0.330
18.5	25.6	38.3	194.8	0.287	0.229

physical parameters for the pond are given in Table 2.3. The measured meteorological variables and spray performance (in terms of efficiency η) are shown in Table 2.4, as well as the NRC model predictions. Figure 2.10 shows the predicted efficiency versus the measured efficiency.

The scatter is much smaller than in the comparison of the model to the Canadys data. This is probably an indication that the experiments were conducted more carefully at Rancho Seco. The NRC model clearly underpredicts the efficiency, and should, therefore, be considered conservative for temperature computations.

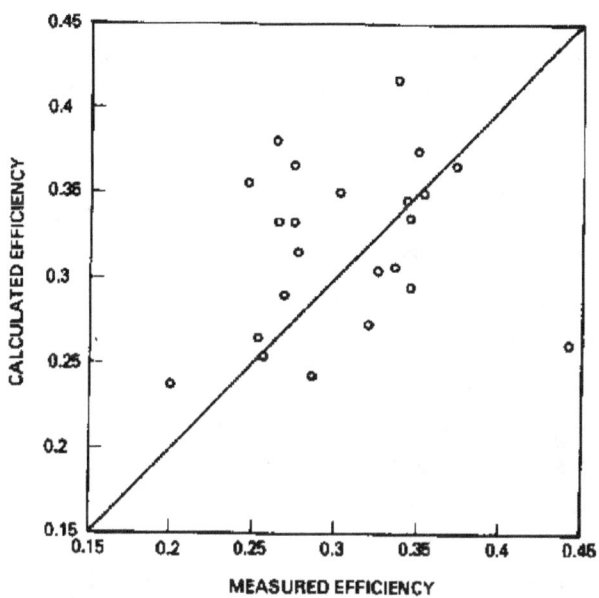

Figure 2.8 Measured and predicted performance of Canadys pond, complete spray model (high windspeed)

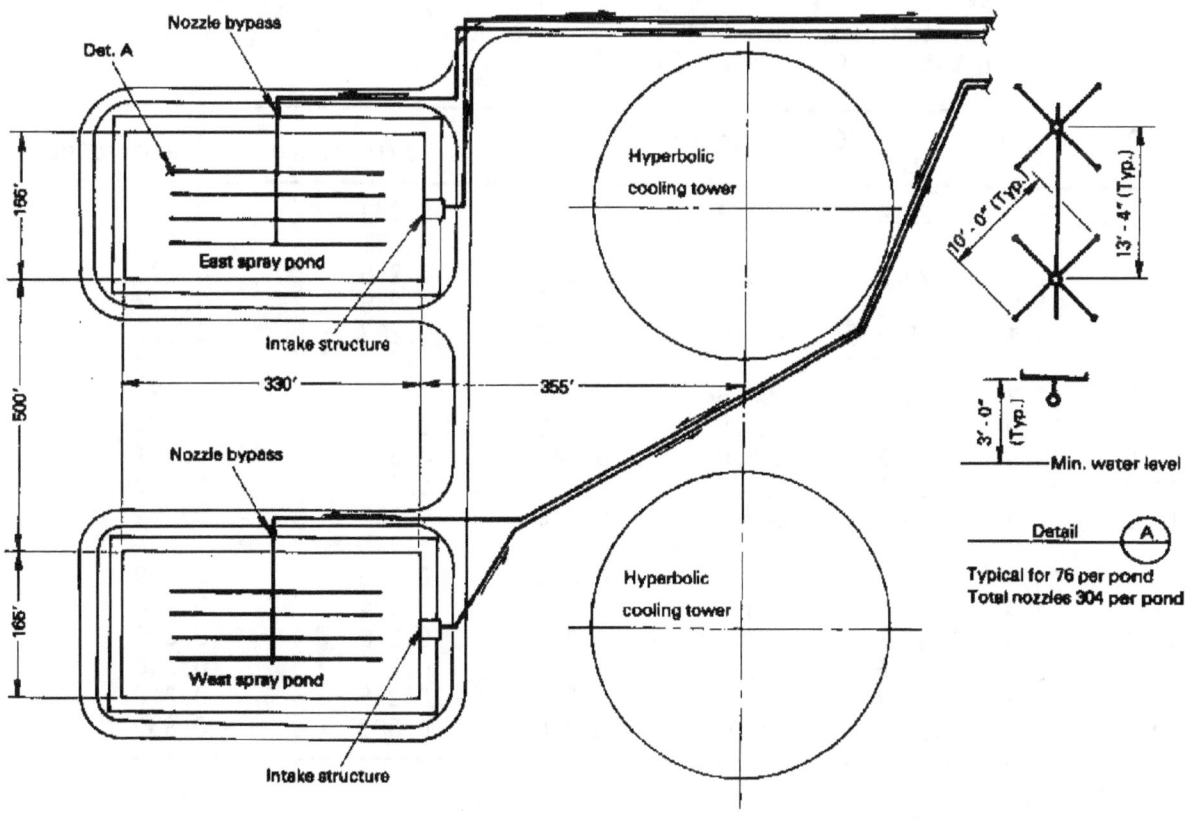

Figure 2.9 Rancho Seco spray-cooling ponds (not to scale)

Table 2.3 Physical Characteristics of
Rancho Seco Spray Pond Used
in Spray-Field Model

Variable	Measurement
Length of spray field	84.8 m
Width of spray field	35.1 m
Height of spray field	3.66 m
Initial drop velocity	6.67 m/sec
Angle of drop with respect to horizon	76°
Height of nozzles from water surface	1.52 m
Barometric pressure	29.92 in. Hg
Flowrate through all nozzles	1590 liters/sec

Table 2.4 Measured Atmospheric Parameters and Spray Efficiency, and Efficiency
Predicted From Combined High-Windspeed and Low-Windspeed Model With
and Without Drag Terms Included

T_w,°C	T_A,°C	T_{HOT},°C	w, cm/sec	$\eta_{measured}$	$\eta_{calculated}$*	$\eta_{calculated}$**
16.1	27.5	26.6	581.8	0.417	0.383	0.415
16.4	27.2	26.7	558.8	0.475	0.381	0.414
10.6	12.8	25.2	236.9	0.325	0.259	0.276
9.2	11.1	25.2	44.7	0.288	0.248	0.277
13.6	18.3	25.3	268.2	0.309	0.287	0.307
14.2	21.7	25.9	290.6	0.355	0.303	0.324
22.4	35.0	26.7	312.9	0.389	0.398	0.423
20.9	33.9	27.3	295.0	0.343	0.368	0.391
19.2	29.8	27.1	375.5	0.458	0.373	0.400
16.1	22.4	26.8	169.9	0.345	0.256	0.261
15.7	20.7	26.5	169.9	0.285	0.250	0.270
12.3	14.4	38.6	44.7	0.352	0.324	0.350
11.7	13.9	37.8	71.5	0.362	0.318	0.348
11.1	13.3	36.6	58.1	0.344	0.310	0.340
9.4	11.7	38.7	44.7	0.345	0.315	0.340
8.9	10.6	36.3	17.9	0.346	0.302	0.330

*With drag terms.
**Without drag terms.

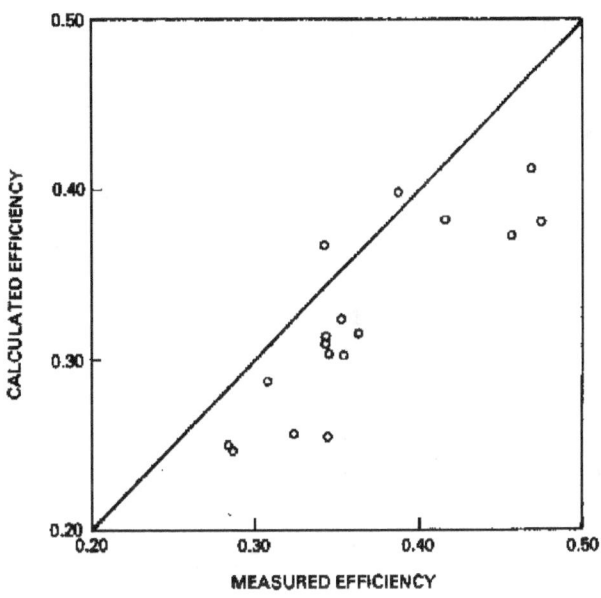

Figure 2.10 Measured and predicted performance of Rancho Seco pond, complete spray model (with drag terms)

2.4 Simplifying Assumptions for Performance Models

The microscale model of the falling drop has been formulated in considerable detail. The possibility of simplifying this facet of the model is explored by starting with a more complete numerical solution of the falling drop and comparing the results to simplified versions of the model (for example, by eliminating one or more terms from the equations). If the results using the simplified model can be shown to be acceptable, substantial reductions in computing time can be realized. In addition, troublesome aspects of the computations can be eliminated if it can be shown that their effects on the performance of the model are negligible.

2.4.1 Simplification for Average Drop Diameter

The motion of the drop and its heat, mass, and momentum-transfer properties depend strongly on its diameter. The drop-diameter distribution in 10 divisions for the Spraco 1751A nozzle is illustrated in Table 2.5. As suggested in

2-25

Table 2.5 Drop-Diameter Distribution for
Spraco 1751A Nozzle

Diameter, cm.	Percent of total	Cumulative volume, %
0.075	10	10
0.12	10	20
0.15	10	30
0.184	10	40
0.22	10	50
0.245	10	60
0.27	10	70
0.31	10	80
0.36	10	90
0.45	10	100

Source: Summarized from Reference 3.

Section 2.1, the heat, mass, and momentum transfers in any segment of the pond
can be found by integrating the contributions over the range of drop diameters.
In practice, the drop-diameter distribution may be broken up into j diameter
ranges and the contribution from each diameter range summed to get the average.
For example, the average drop temperature T is:

$$T = \sum_{i=1}^{j} f_i T_i \quad °C \qquad (2.49)$$

The problem with this approach is that there must be a solution of the equations
for each of the j drop diameters. If instead, a single average drop diameter
could be found, which gave the same results as the summation of the results
for the j individual drop diameters, the computational effort would be reduced
by a factor of about 1/j.

It is not obvious that an average drop diameter exists which would consistently
duplicate the performance of the spray model using the distributed drop-diameter
formulation. In order to test the theory that an acceptable mean diameter could
be used, the HWS and LWS models were run over a wide range of conditions, using
an observed drop-diameter distribution. The resulting performances were then
compared to the results of the HWS and LWS models using a single drop diameter
over the same range of conditions.

In all cases for which it was tested, it appears that a single average drop
diameter <u>can</u> be chosen to very nearly represent the drop-diameter distribution
over a wide range of operation for both the LWS and HWS models. Figures 2.11
and 2.12 illustrate that for the HWS and LWS models, the "average" drop diameter
which gives results closest to the distributed drop for the Spraco 1751 nozzle
is about 0.208 cm for the LWS model and 0.196 cm to 0.202 cm for the HWS model.
Figure 2.13 demonstrates for the HWS model how closely the "average drop
diameter" model compares to the "distributed drop diameter" model.

2.4.1.1 Estimating the Average Drop Diameter

The average diameter illustrated above was determined by experimentation with
the model on a single drop-diameter distribution and spray-pond configuration.
It is difficult to generalize how one would estimate the average drop diameter
under completely general conditions, except to illustrate how well the empirically determined average diameter works over a wide range of conditions for both
the HWS and LWS spray-pond models.

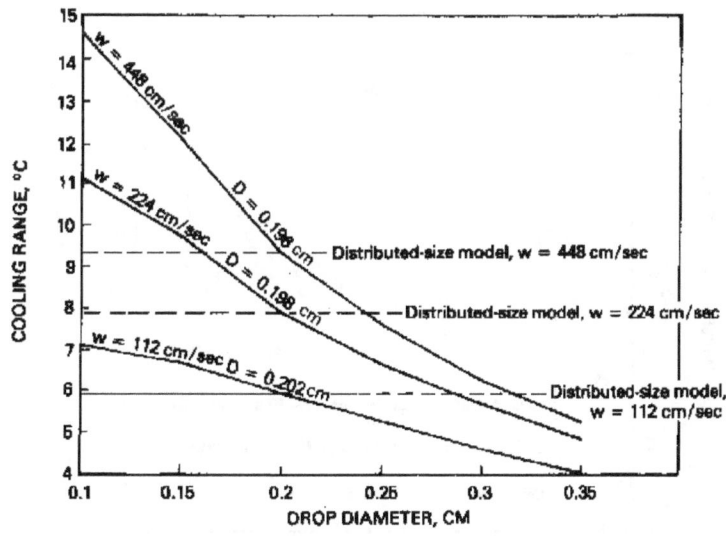

Figure 2.11   Determination of "average drop diameter" for high-windspeed model

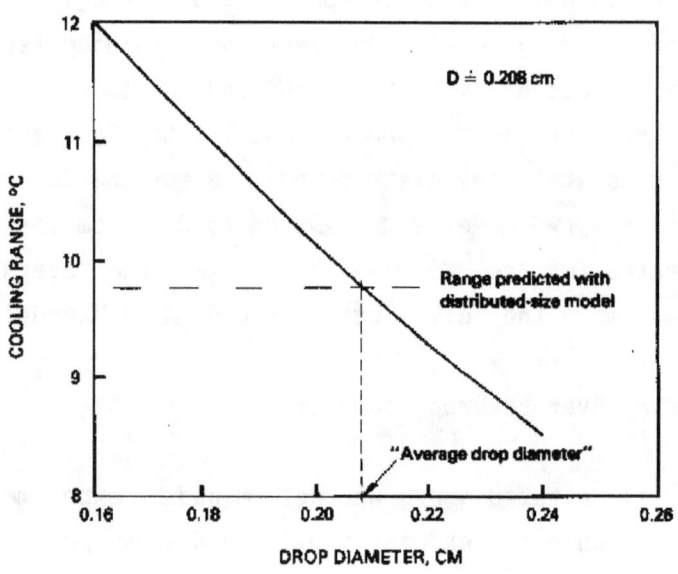

Figure 2.12 Determination of "average drop diameter" for low-windspeed model

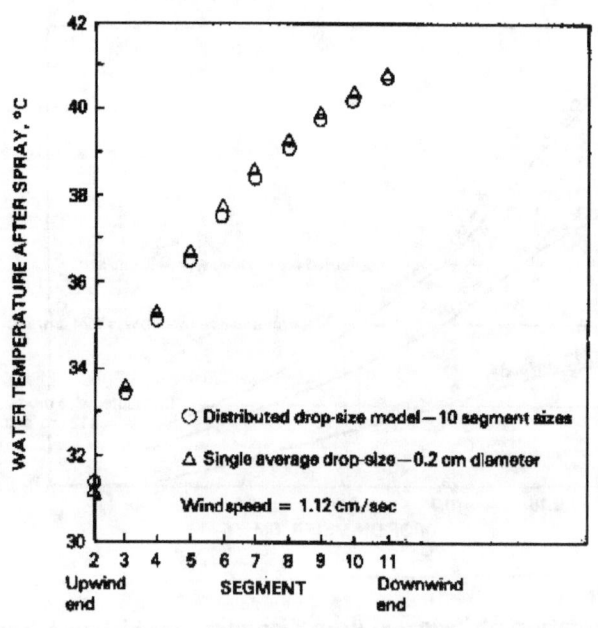

Figure 2.13 Performance of high-windspeed model for mean and distributed drop diameter

A formula which has been developed on physical principles to represent the mean drop diameter in heat and mass transfer from drops is the "Sauter" mean (Ref. 12), which is based on an area-weighted mean volume:

$$D_3 = \frac{\int_0^\infty D^3 F(D)}{\int_0^\infty D^2 F(D)} \doteq \frac{\sum_{i=1}^{j} D_i^3 f_i}{\sum_{i=1}^{j} D_i^2 f_i} \qquad (2.50)$$

where

$F(D)$ = probability density function (PDF) for the drop-diameter distribution = differential of cumulative distribution function (CDF)

D = drop diameter, cm

D_i = drop diameter in diameter range i, cm

f_i = fraction of drops by mass in diameter range i.

For the Spraco 1751 nozzle drop-diameter distribution shown in Table 2.5, the Sauter mean calculated by the discrete form of Eq. 2.50 is $D_3 = 0.339$ cm. This is somewhat larger than the mean diameter from 0.2 cm to 0.208 cm, which was determined to give the best agreement with the distributed diameter model.

Use of the Sauter mean would result in a lower cooling efficiency than would be predicted by the "correct" method using the distributed drop diameters.

It is possible to define a general class of mean diameters D_n:

$$D_n = \frac{\int_0^\infty D^n F(D)}{\int_0^\infty D^{(n-1)} F(D)} \doteq \frac{\sum_{i=1}^{j} D_i^n f_i}{\sum_{i=1}^{j} D_i^{(n-1)} f_i} \qquad (2.51)$$

For example, the Sauter mean would be called D_3. Figure 2.14 shows the nth order mean diameter D_n calculated from the discrete form of Eq. 2.51 versus the order n for the distribution shown in Table 2.5. The order of the mean which yields the empirically determined mean diameter of 0.208 cm is about n = +0.45. Since larger drop diameters are conservative, we will arbitrarily pick an order of the mean n = 0.5, which gives a mean diameter of 0.211 cm. Equation 2.51 for n = 0.5 reduces to:

$$D_{\frac{1}{2}} = \frac{\int_0^\infty \sqrt{D} \; F(D)}{\int_0^\infty \frac{F(D)}{\sqrt{D}}} \doteq \frac{\sum_{i=1}^{j} \sqrt{D_i} \; f_i}{\sum_{i=1}^{j} \frac{f_i}{\sqrt{D_i}}} \tag{2.52}$$

This is the suggested diameter to be used in the HWS and LWS performance models.

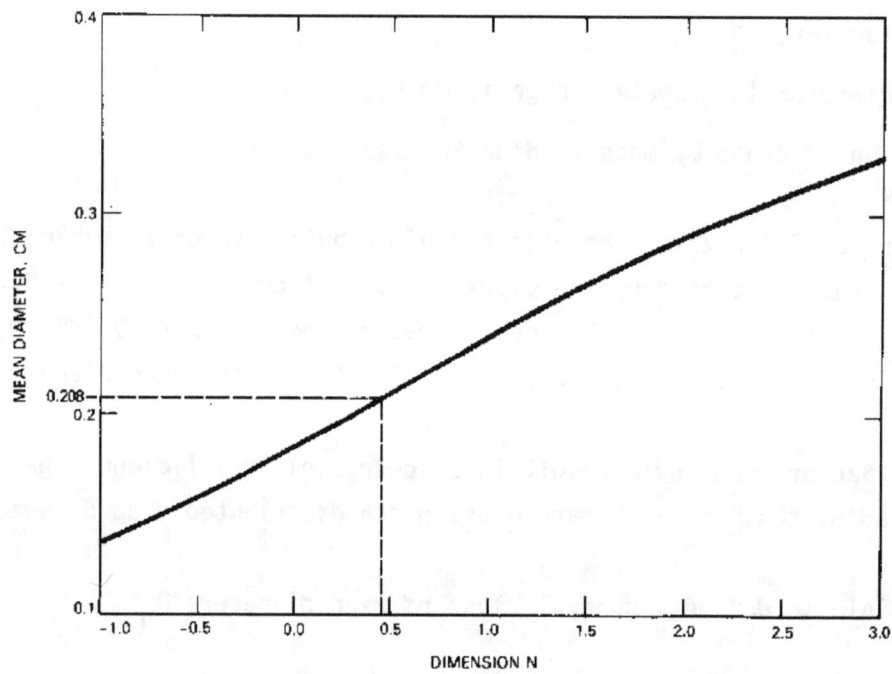

Figure 2.14   Determination of order of the mean for Spraco 1751A
drop-diameter distribution

2.4.2 Effect of Drag on Performance Models

Including drag on the falling drops introduces several complications to the model, most notably:

(1) The drag term makes the equations of motion for the drop (Eqs. 2.1 and 2.2) nonlinear, requiring a numerical integration solution. By eliminating the drag term, the motion of the drop can be described analytically.

(2) On the LWS model, the net downward drag of the drops is a destabilizing influence on the iterative solution, especially at low heat loads.

For these and other reasons, it would be highly desirable to eliminate the drag term from Eqs. 2.1 and 2.2. The effect of eliminating the drag term from the HWS and LWS models was tested for a typical spray-pond configuration over a wide range of heat loading and atmospheric conditions. The following is a discussion of the various effects resulting from drag elimination.

2.4.2.1 Microscale Submodel

Eliminating the drag terms in Eqs. 2.1 and 2.2 has two effects:

(1) The time of flight is shortened.

(2) The rate of heat and mass transfer is increased because the average drop velocity is higher.

These two phenomena counteract each other to a certain extent, but the net effect is that the falling drops are predicted to experience more cooling and evaporation once drag is eliminated.

2.4.2.2 Macroscale Model

Eliminating the drag term increases the efficiencies predicted by both the HWS and LWS models. In addition, it increases the stability of the iterative solution in the LWS model. Table 2.4 shows the predicted efficiencies for the HWS and LWS models with and without drag over a range of heat and meteorological conditions for the Rancho Seco spray-pond test. Figure 2.15 compares the combined HWS-LWS "no-drag" model results (choosing the higher η of the two) with the Rancho Seco test data. The model-prototype agreement is good, and the no-drag model results are still conservatively low. In fact, agreement is better <u>without</u> drag than <u>with</u> drag, because the elimination of drag raises the predicted efficiency.

On the basis of the good agreement to data shown by the model and the improvement in stability of the LWS model, the drag term can be eliminated for typical spray-pond applications. This would not be a correct assumption for certain oriented spray configurations that are designed to induce lateral air flows (Ref. 7). In those cases, the effects of drag would have to be included. In addition, drag cannot be neglected in the drift-loss model described in the next section, since the smaller drop diameters which are most prone to drift, are strongly affected by drag.

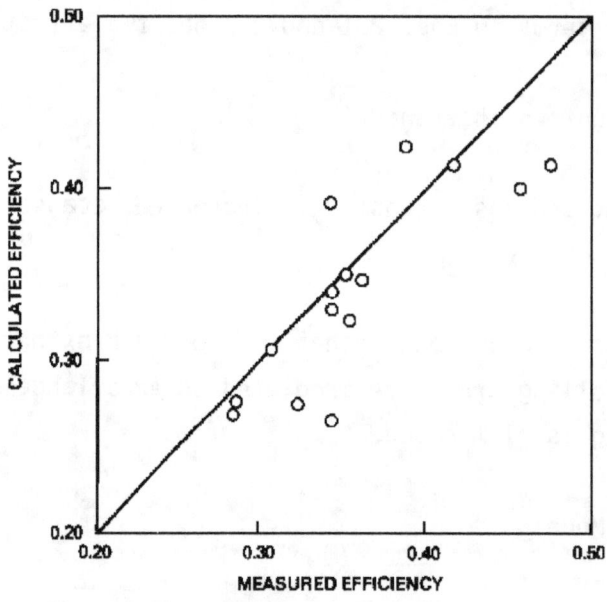

Figure 2.15 Comparison of NRC model with Rancho Seco data
for "no-drag" model

3 DRIFT-LOSS MODEL

A fraction of the water droplets sprayed from the nozzles will be lost because
they are physically carried by the wind beyond the pond borders. This "drift"
loss can be estimated by means of a mathematical model showing the trajectory
of droplets in a wind field and where the droplets fall in relation to the
borders of the pond. Drift losses are generally small compared with evaporative
losses.

3.1 Model Assumptions

The model is formulated for a spray pond of conventional design, with the
Spraco 1751A nozzle operating at the recommended pressure and height. The
trajectories of drops leaving the spray nozzles are simulated using a ballistics
approach, in a similar manner as the "microscale" submodel of Section 2.1, but
for 21 drop diameters which represent the drop-diameter distribution of the
Spraco 1751 nozzle rather than the 10 drop diameters used in performance models.
The equations in Section 2.1.1 apply exactly. No interaction of drops is
presumed. It is likely that this is a conservative assumption, since small
drops in some cases would collide to form larger drops which are less prone to
be carried by wind.

The process of drop formation is complicated. Water will generally leave the
nozzles in a continuous stream. Once the stream is airborne, forces of surface
tension tend to cause the breakup of the stream into drops of varying sizes.
Aerodynamic forces may also cause the larger drop diameters to become unstable
and break up into smaller, more-stable drops. In every case, the breaking apart
of larger drops into smaller drops causes the formation of one or more very
small particles separate from the two major components of the fission. The
drop-diameter distribution is not only a function of the type of spray nozzle and
pressure, but of the distance from the nozzle, since the breakup into smaller
drops occurs along the entire path.

If the assumption were made that all particles were already formed leaving the nozzle, drift loss would be underestimated. This is because the smallest particles most prone to drift also have small momentum, and would not be predicted to attain a very great height with respect to the nozzle. The most conservative assumption in this case would be that all droplets are formed at the apogee of the trajectory of the largest drop diameter, even though many small drops form close to the nozzle.

The buoyancy of the heated, humidified air in a heavily loaded spray pond could cause an updraft on the order of tens to hundreds of centimeters per second during low wind conditions. A single value of updraft velocity is chosen and inputted to represent an average for the 30-day period of an accident. The default value is 50 cm/sec.

3.1.1 Ballistics Model for a Drop

The model for the flight of the drops is the same as that developed in Section 2.1.1 and will not be repeated. It should be noted, however, that more emphasis is placed on the trajectory of small drops in the drift model; these are relatively less important for the spray-heat-loss calculations of Section 2. Therefore, a finer drop-diameter distribution is needed. The default drop-diameter distribution used for the drift model is shown in Table 3.1.

3.1.2 Initial and Boundary Conditions

The Spraco 1751 under a design pressure of 7 psig demonstrates a nozzle velocity of about 24 ft/sec. The spray would form a cone of water with an average angle of 58° from the horizontal. In calm conditions, the sprayed water forms an "umbrella" of about 12 ft in height and up to 16 ft in radius when the nozzle is 5 ft above the water surface according to Spraco promotional literature (Ref. 11).

Under the influence of wind, the spray umbrella is distorted. The circular pattern of droplets falling on the water surface is shifted downwind. The apogee of the drops is decreased in the upwind direction and increased in the

Table 3.1 Default Drop-Diameter Distribution
for Spraco Nozzle 1751A for Use in
Drift Model

Diameter, microns	Percent of total	Cumulative volume, %
200	0.05	0.05
260	0.05	0.1
300	0.05	0.2
330	0.1	0.3
365	0.1	0.4
400	0.1	0.5
425	0.2	0.7
460	0.3	1.0
520	0.4	1.4
580	0.6	2.0
640	2.0	4.0
855	3.0	7.0
1000	3.0	10.0
1190	5.0	15.0
1340	5.0	20.0
1650	10.0	30.0
2000	10.0	40.0
2290	10.0	50.0
2800	20.0	70.0
3600	15.0	85.0
4000	15.0	100.0

Source: See Reference 3

downwind direction. The smaller drop diameters would naturally be affected more than the larger ones. All drops of the same diameter would fall roughly in a circular pattern, however. This last assumption simplifies the analysis somewhat, because the diameter of the circular pattern for a particular drop diameter can be determined from just the straight upwind and straight downwind trajectories of the spray.

The starting point for the trajectory computations for all drop diameters is conservatively chosen as the apogee of the largest drop diameter, for reasons previously discussed.

The velocities and vertical and horizontal coordinates of both the upwind and downwind apogees for the largest drop diameter are calculated for a range of windspeeds and stored. These stored values are then used as the initial conditions for each windspeed in Eqs. 2.1 and 2.2 for the range of 22 drop diameters representing the spray-diameter distribution.

The circular patterns for each windspeed and each drop diameter, which are predicted from the drop ballistics, are used subsequently to predict the fraction of water passing beyond the boundaries of the pond. A drop is assumed to be lost if it does not fall on the pond surface. No allowance is made for runoff from the berms back into the pond.

The critical pond boundary is a straight line, arbitrarily oriented to be closest to the greatest number of nozzles in the downwind direction, as illustrated in Figure 3.1. The distance of the nozzles, or group of nozzles, equidistant from this line and the fraction of water in each group is specified. The part of the circular pattern for each drop diameter and wind falling outside of the critical pond boundary is then calculated for each nozzle or group of nozzles, which is the drift loss.

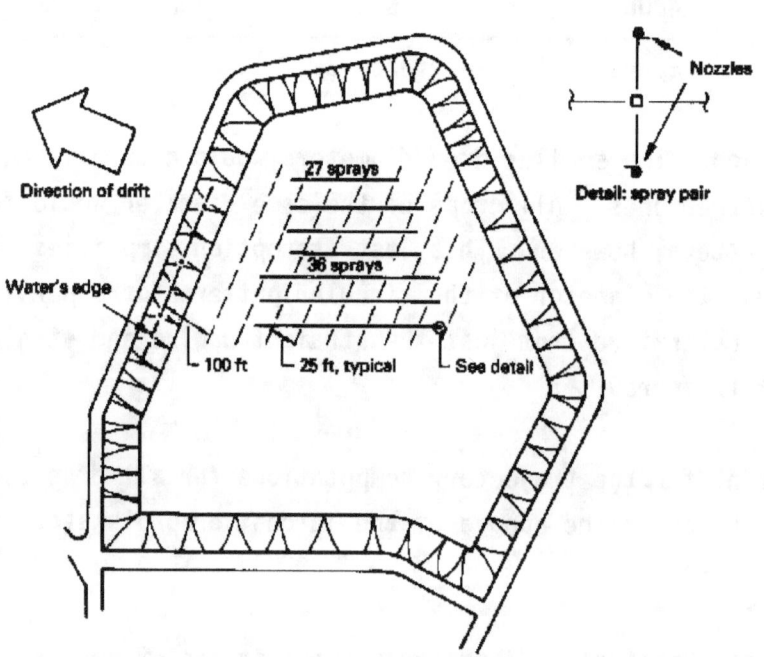

Figure 3.1 Typical layout of pond sprays and determination of critical pond boundary

3.2 Model Validation

3.2.1 Rancho Seco Data

Only limited field data are available on actual spray ponds with which the drift model can be validated.

The Rancho Seco drift-loss data were collected during an operational test of the spray-pond system required by NRC for the licensing of the plant (Ref. 10). Pond inventory and windspeed measurements were made during the test period, and then used to estimate the fraction of sprayed water lost versus windspeeds, which were typically from 0 to 15 mph.

To account for evaporation under zero heat load, the investigators conservatively estimated the drift loss by subtracting the water-loss rate at zero windspeed from the rest of the data. They erroneously assumed that evaporation from the pond and sprays would be independent of windspeed. Actually, evaporation from both the pond surface and spray increases directly with the windspeed. They therefore overestimated the water loss due to drift by neglecting the additional evaporation from the sprays. The water-loss data for the no-heat-load run (No. 4) of the Rancho Seco test are plotted versus windspeed in Figure 3.2.

The results of the drift-loss model cannot be directly compared to the prototype data in Figure 3.2 without first estimating the evaporative losses of the sprays, even without external heat loads. Unfortunately, there were no meteorological data other than windspeeds readily available from the no-heat-load test. On the basis of data that were available from other tests in the series, however, two combinations of wet-bulb/dry-bulb temperature values were estimated, which probably bound the range of meteorological conditions other than wind during the test.

The correction factor for evaporation of the sprays was computed directly from the high windspeed (HWS) performance model described in Section 2.2.1, which was run under no-heat-load conditions for a range of windspeeds. The sprayed

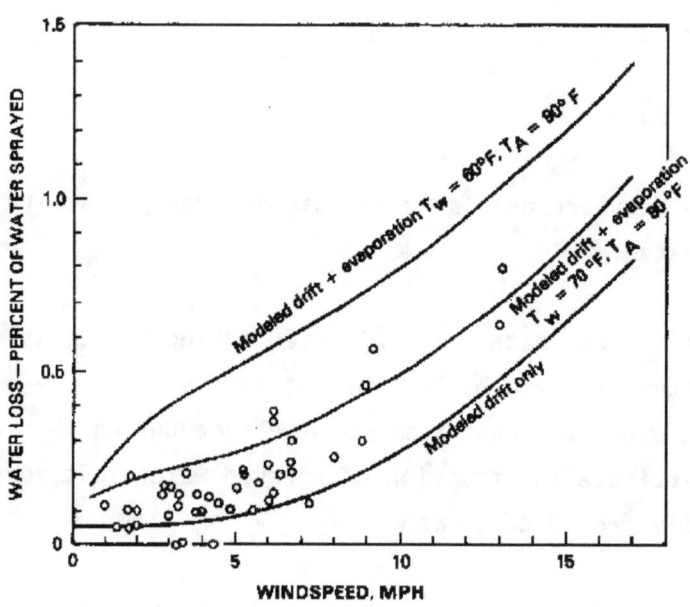

Figure 3.2   Measured and modeled water loss from Rancho Seco test 4

temperature T_{HOT} was forced to be equal to the temperature after spraying T by running the program iteratively until convergence.   Two cases were run:

(1)   Wet-bulb temperature = 70°F
        Dry-bulb temperature = 80°F

(2)   Wet-bulb temperature = 60°F
        Dry-bulb temperature = 90°F

Additionally, a correction factor has been added to account for the relatively minor contribution of heat to the spray pond from solar radiation.   Mean daily solar radiation during May is about 2,450 Btu (ft² day) (Ref. 13).   The surface area of the full pond is about 66,470 ft².   If 80% of this added heat is lost through evaporation, it would correspond to a water loss of 0.0239 ft³/sec. The quantity of sprayed water during the test was about 35.4 ft³/sec, which means that water evaporated because of solar heat load would be about 0.067% of the volume sprayed.

The water loss during the no-heat-load test is, therefore, calculated to be:

$$W_1 = \text{drift loss} + \left(\begin{array}{l}\text{evaporative loss} \\ \text{for no-heat load}\end{array}\right) + \left(\begin{array}{l}\text{solar heat} \\ \text{load correction}\end{array}\right) \qquad (3.1)$$

Water loss versus windspeed is plotted in Figure 3.2 for the two assumed meteorologic conditions, along with data from the no-heat test at Rancho Seco. The model appears to conservatively follow the field data on water loss, although it must be recognized that no detailed meteorological conditions were readily available for this comparison.

3.2.2 Validity of Drift-Loss Model

The drift-loss model presented here has been shown to perform acceptably well when compared to the limited field data available and incorporates a number of conservatisms in its formulation. Greater emphasis on the drift-loss model is probably not warranted, since the total quantity of water lost to drift is generally much smaller than water lost to evaporation. Drift loss may exceed evaporation momentarily during high winds but it is unlikely that these conditions could be sustained for a sufficient length of time to change this conclusion.

4 POND MODEL

The pond model is used to calculate the temperature and water loss from the pond. It combines the model of heat and mass loss from the sprays, the model of circulation and heat retention in the mass of water in the pond, and additional heat and mass transfer from the surface of the pond. The pond model developed here is similar to the mixed-tank model of NUREG-0693 (Ref. 14).

A typical spray pond differs from surface-cooling ponds by having smaller volume and surface areas. The rates of heat and water loss from the sprays to heat and water loss from the pond surface is generally high.

The heat and water loss from the pond surface may in most cases be considered a secondary effect with regard to the sprays. In addition, the small volumes of the ponds relative to the water circulation through them diminishes the effects of such phenomena as thermal stratification, which are of importance in surface-cooling ponds (Ref. 15). For this reason, the modeling of the balance of the pond other than the sprays is fairly straightforward and simple. The "mixed tank" model of the pond assumes total mixing of all water throughout the volume of the pond. It must be noted, however, that some spray ponds may have a relatively large surface area and volume, or the sprays may be operated only intermittently. In these cases, surface-heat transfer and the effects of stratification may take on greater importance than in a typical spray-pond situation. The effects of "short-circuiting" of pond water are not nearly as important in typical spray ponds as they could be in surface-cooling ponds.

The mixed-tank model depicted in Figure 4.1 presumes that the heated effluent is instantaneously and uniformly mixed throughout the volume of the tank, and that the water in the tank is uniform in temperature. Atmospheric-heat transfer from the surface is related to the pond-surface temperature.

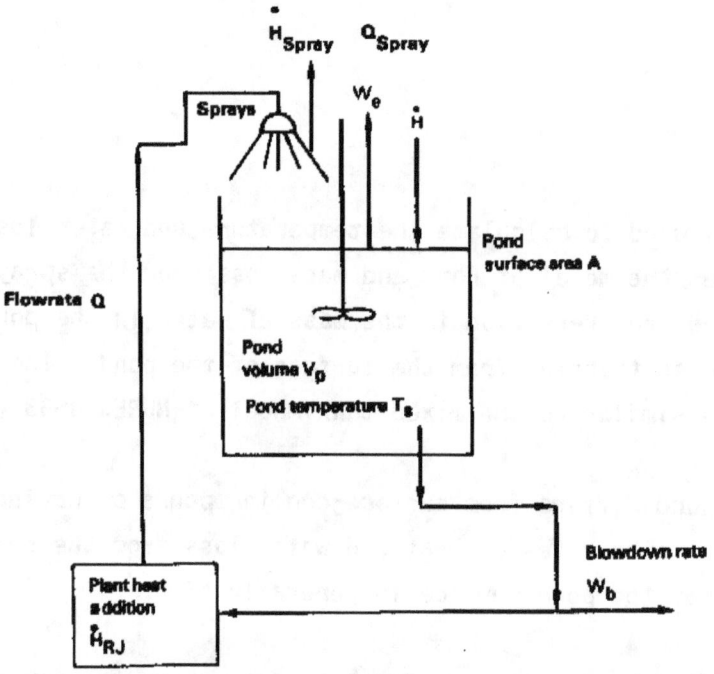

Figure 4.1 Mixed-tank model

4.1. Heat Balance

A heat-and-mass balance can be formulated for the mixed-tank model. The terms of the heat balance are:

4.1.1 Heat Load Into Ponds

$$\text{Heat in} = \dot{H}_{RJ} \qquad \text{Btu/(ft}^2 \text{ day)} \tag{4.1}$$

4.1.2 Heat Out From Surface

A relation for the rate of net heat flow across the surface of the pond can be developed through consideration of each heat source and heat loss. The net rate of heat flow $\dot{H}$ into the pond is:

$$\dot{H} = \dot{H}_{SN} + \dot{H}_{AN} - \dot{H}_{BR} - \dot{H}_{E} - \dot{H}_{C} \qquad \text{Btu/(ft}^2 \text{ day)} \tag{4.2}$$

where

$\dot{H}_{SN}$ = net rate of shortwave solar radiation entering the pond, measured directly, Btu/(ft² day)

$\dot{H}_{AN}$ = net rate of longwave atmospheric radiation entering the pond, measured directly, Btu/(ft² day)

$\dot{H}_{BR}$ = net rate of back radiation leaving the pond surface, Btu/(ft² day)

$\dot{H}_{E}$ = net rate of heat loss attributable to evaporation, Btu/(ft² day)

$\dot{H}_{C}$ = net rate of heat flow from the pond attributable to conduction and convection, Btu/(ft² day)

The relationships are illustrated graphically in Figure 4.2.

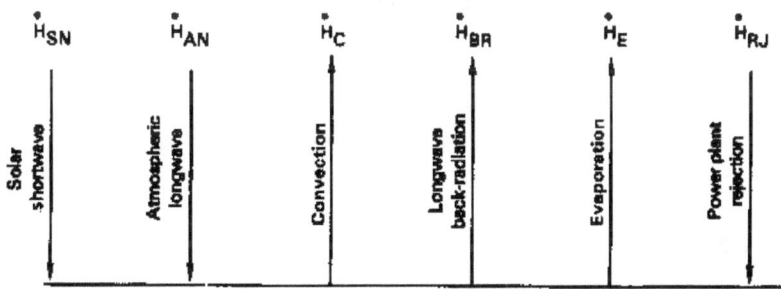

Figure.4.2 Heat loads on the surface of a pond

The net atmospheric radiation term can be approximated using air temperature T_A and cloud cover C. Ryan and Harleman (Ref. 16) develop the following formula for $\dot{H}_{AN}$:

$$\dot{H}_{AN} = 1.2 \times 10^{-13}\left(T_A + 460\right)^6\left(1 + 0.17C^2\right) \qquad \text{Btu/(ft}^2 \text{ day)} \qquad (4.3)$$

The back radiation term may be expressed using the relation for radiation from a black body (Ref. 17):

$$\dot{H}_{BR} = 4.026 \times 10^{-8}\left(460 + T_s\right)^4 \qquad \text{Btu/(ft}^2 \text{ day)} \qquad (4.4)$$

4-3

The evaporative-heat-transfer component is a function of surface temperature at atmospheric temperature and humidity:

$$\dot{H}_E = \left(e_s - e_a\right)F(w) \qquad \text{Btu/(ft}^2\text{ day)} \tag{4.5}$$

where

e_s = vapor pressure of water at the pond-surface temperature, mm Hg

e_a = partial pressure of water vapor in the air (that is, the vapor pressure of water at the dewpoint), mm Hg

$F(w)$ = wind function

A semiempirical wind function is proposed by Ryan and Harleman (Ref. 16) which agrees well with field data on large ponds:

$$F(w) = \left[22.4 \times \left(\Delta T_v\right)^{1/3} + 14w\right] \tag{4.6}$$

where

w = windspeed, mph

ΔT_v = "virtual" temperature difference between the pond surface water and air above the pond, rewritten:

$$\Delta T_v = \frac{T_s + 460}{1 - \dfrac{0.378 \times e_s}{p}} - \frac{T_A + 460}{1 - \dfrac{0.378 \times e_a}{p}} \tag{4.7}$$

and p = atmospheric pressure, mm Hg

The net rate of heat transfer from the pond attributable to conduction and convection, H_c, is also a function of the pond surface and atmospheric humidity and temperature (Ref. 16):

$$H_c = 0.26 \times \left(T - T_A\right) \times F(w) \tag{4.8}$$

4.1.3 Heat Out in Blowdown or Leakage Stream

With reference to the pond temperature T, heat loss from blowdown is by definition zero:

$$q_b = W_b \rho C_p (T - T) \equiv 0 \quad \text{Btu/hr} \tag{4.9}$$

where

W_b = flowrate of the blowdown or leakage stream, ft^3/hr

and ρ and C_p are as previously defined.

4.1.4 Heat Rejected by Sprays

$$H_{spray} = Q \rho C_p R_c \quad \text{Btu/(ft}^2 \text{ day)} \tag{4.10}$$

where

R_c = cooling range of the sprays determined from either the HWS-LWS model or the regression equations

and Q, ρ, and C_p are as previously defined.

Combining all heat inputs to and outputs from the pond, and using the relationship relating temperature to heat, the following relationship is obtained:

$$\frac{dT}{dt} = \frac{\dot{H}_{RJ} - \dot{H} - \dot{H}_{spray}}{\rho C_p V_p} \quad \text{°F/hr} \tag{4.11}$$

where

V_p = pond volume in cubic feet

and all other elements of the equation are as previously defined.

Note that there is no provision for a makeup stream in either the heat or mass balance, since Regulatory Guide 1.27 specifically denies makeup during the operation of a UHS pond.

4.2 Mass Balance

The mass balance on the pond includes evaporative loss from the surface, drift, and blowdown or leakage. The terms of the mass balance are:

Blowdown or leakage flow = W_b, ft³/hr
Evaporative loss from surface = W_e, ft³/hr

$$W_e = \frac{\dot{H}_E}{\rho\lambda} \qquad (4.12)$$

where

λ = heat of vaporization of water, Btu/lb

ρ = density of water, lb/ft³

and $\dot{H}_E$ is defined by Eq. 4.5.

Combining all terms of the mass balance yields the expression:

$$\frac{dV}{dt} = -W_b - \frac{\dot{H}_E}{\rho\lambda} - W_{drift} - W_{spray} \qquad (4.13)$$

where

W_{drift} = drift loss

W_{spray} = rate of water evaporated from all drops in the spray field, ft³/hr

determined from the evaporative heat-transfer component of Eq. 2.7.

5 DATA-SCREENING METHODOLOGY

In this section, a method is developed with which long-term weather records can be screened to find the period in which the spray-pond temperature or water loss will be maximized.

5.1 Development of Method

The "equilibrium temperature" heat-transfer approach is used in a method that decouples the plant-heat-input effects from environmental effects on the pond. The temperature of the pond, T_s, may be determined by the solution of the differential equation for the mixed-tank model:

$$\frac{dT_s}{dt} = \frac{\dot{H}}{\rho C_p V_p} + \frac{Q\eta}{V_p}\left(T_s + \frac{\dot{H}_{RJ}A}{\rho C_p Q} - T_W\right) \tag{5.1}$$

where

V_p = pond volume, ft^3

A = pond surface area, ft^2

T_W = wet-bulb temperature, °F

and all other elements are as previously defined.

For the purpose of developing the model, V_p and η are temporarily assumed to be constant. The "equilibrium temperature" E (Ref. 17) is a useful invention at this point in the model development. The rate of atmospheric-heat transfer can be assumed to be proportional to the difference between the pond temperature and the equilibrium temperature:

$$\dot{H} = KA\left(T_s - E\right) \tag{5.2}$$

where

K = equilibrium-heat-transfer coefficient, $Btu/(ft^2 hr °F)$

If we further assume that K is a constant, Eq. 5.1 will be linear with respect to T_s, and it will be possible to consider that the pond temperature is the sum of the pond "ambient" temperature T_s' and an "excess" temperature θ:

$$T_s = T_s' + \theta \tag{5.3}$$

T_s' would be determined by the solution of Eq. 5.1 for a steady heat load $\dot{H}_{RJ,0}$:

$$\frac{dT_s'}{dt} = \frac{AK}{\rho C_p V_p} (T_s' - E) - \frac{Q\eta}{V_p} \left(T_s' + \frac{\dot{H}_{RJ,0} A}{\rho C_p W} - T_W \right) + \frac{\dot{H}_{RJ,0} A}{\rho C_p V_p} \tag{5.4}$$

where

$\dot{H}_{RJ,0}$ = steady-state heat load, $Btu/(ft^2 \ day)$

and all other values are as previously defined.

Subtracting Eq. 5.4 from Eq. 5.1 gives the differential equation for excess temperature:

$$\frac{d\theta}{dt} = \frac{AK}{\rho C_p V_p} \theta + \left(\frac{\dot{H}_{RJ} - \dot{H}_{RJ,0}}{\rho C_p V_p} \right) - \frac{Q\eta}{V_p} \left(\theta + \frac{\dot{H}_{RJ} - \dot{H}_{RJ,0}}{\rho C_p Q} - T_W \right) \tag{5.5}$$

The determination of pond temperature has, therefore, been separated into two simpler problems, because now the ambient and excess pond temperatures can be determined independently from one another. The excess temperature θ does not depend on the meteorological record, so it can be solved directly from Eq. 5.5

using the plant-heat-rejection rate. The pond ambient temperature T'_s does not depend on the heat rejection from the plant, so it can be calculated from Eq. 5.4 using only the long-term meteorological record. The peak pond temperature can, therefore, be found by summing (superimposing) the peak of T'_s and θ:

$$\left(T_s\right)_{peak} = \left(T'_s\right)_{peak} + \theta_{peak} \tag{5.6}$$

Unfortunately, the basic premise that Eq. 5.1 is linear is incorrect. Both K, E, and η are functions of T_s and atmospheric variables. In addition, the pond volume V_p will change as water on the pond is lost as a result of seepage, drift, and evaporation. (Makeup water is assumed to be unavailable during the operation of the pond.) The function of the procedure outlined above is to identify the timing of the maximum ambient and maximum excess temperatures so that more accurate computation can be performed in which the spray-pond temperature is determined directly. Since the heat- and mass-transfer relationships are nonlinear with respect to pond and spray temperature, temperature calculations may be different from those used in the screening. There are, however, no firm guarantees that the optimal starting time for peak temperature will necessarily be found by this procedure. A series of model runs spaced several hours apart, over the length of the data record, is an alternative method of determining the optimal timing.

5.2 Meteorological Inputs to Screening Model

The screening model developed in Section 5.1 requires two types of data: (a) weather data such as wet- and dry-bulb temperatures, dewpoint, windspeed, and atmospheric pressure, which may be obtained from National Weather Service records, and (b) rates of net solar radiation which generally do not exist for long periods of record. A method for synthesizing solar radiation using cloud-cover data has been developed. National Weather Service tapes of "Tape Data Family-14" (TDF-14) are used by the model as a source of temperature, windspeed, and cloud-cover observations. These tapes are available for major observation points throughout the United States.

5.2.1 Solar Radiation

The solar radiation term for the heat-exchange relation must be either taken from direct measurements or estimated. The model estimates hourly solar radiation rates in a three-step process. First, given the latitude of the pond and the time of year, the maximum solar radiation available to the pond for the given day is estimated. Second, this gross figure is fitted to a sinusoidal relation to find the rate of insolation for each hour of daylight. Finally, these hourly rates are modified to take into account the effect of cloud cover.

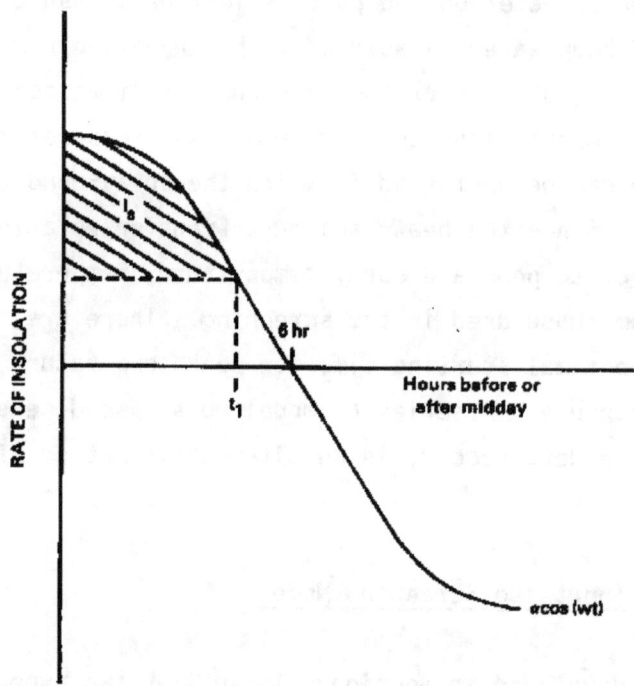

Figure 5.1   Insolation as a function of time

A procedure based on the work of Hamon, Wiess, and Wilson (Ref. 18) is used to estimate the maximum daily solar radiation. This total daily radiation figure is fitted to a sinusoidal function as shown in Figure 5.1. The hourly variation of radiation is:

$$\dot{H}_S(t_0) = 2t_1 \quad \beta\cos\left(\frac{\pi t_0}{12}\right) - \beta\cos\left(\frac{\pi t_1}{12}\right) \quad \text{Btu/(ft}^2 \text{ day)} \tag{5.7}$$

where

$\dot{H}_S$ = gross rate of solar radiation, Btu/(ft^2 day)

t_0 = time of the observation in hours before
or after midday

t_1 = one-half the length of daylight per day, hr

and

$$\beta = \left[\left(\frac{1}{I_s}\right)\left(\frac{\pi}{12}\right)\sin\left(\frac{\pi t_1}{12}\right) - \frac{1}{I_s}t_1\cos\left(\frac{\pi}{12}t_1\right)\right]^{-1}$$

where

I_s = total daily solar radiation, Btu/(ft^2 day)

Solar radiation ultimately reaching the earth's surface is greatly affected by atmospheric conditions, especially cloud cover. The amount of cloud cover, in tenths of the total sky obscured, is available from the data tapes. This information is used in a relationship developed by Wunderlich (Ref. 19) to modify the hourly insolation rates:

$$\dot{H}_{SN} = H_S(1 - 0.65C^2)0.94 \qquad \text{Btu/(ft}^2 \text{ day)} \qquad (5.8)$$

in which 0.94 is a factor which adjusts for the average 6% reflection from the water surface.

5.3 Scanning-Performance Models

In order to determine the design-basis conditions for evaluation of the spray pond, a long-term weather record is searched for key conditions which would predict the highest pond temperature or water-loss rate. Basically, a long-term weather record is searched by using a model which is nearly the same as the model in Section 4 to simulate the performance of a loaded spray pond. The scanning model differs from the model of Section 4 in that the HWS and LWS spray-performance models are not used directly. Using the rigorous performance models for a long (tens of years) simulation would be prohibitively costly and inefficient.

5.3.1 Approximate Spray-Performance Model

The HWS and LWS spray-performance models are steady state. Therefore, they do not depend on any history of input conditions, but predict instantaneous heat rejection and evaporation for a given set of meteorological and heat-load conditions.

If the spray-performance models can be exercised over a wide range of inputted independent meteorological variables, the resulting performances can be formulated into regression models. These regression models can then be used to predict the performance of the sprays for other conditions that are within the ranges of the correlating independent variables. This procedure is much more efficient than using the original models directly.

5.3.2 Functional Dependencies of Spray-Performance Models

Before the regression models are formulated, it is useful to perform numerical experiments using the LWS and HWS models to determine the approximate dependence of predicted performance on the independent variables T_W, T_{HOT}, and w for a typical spray-pond situation. Figures 5.2 through 5.5 show, respectively, the different "spray efficiencies" η of both the HWS and LWS models, that occurred upon variations in wet-bulb temperature, dry-bulb temperature, sprayed-water temperature, and windspeed. The higher of the two predicted efficiencies (LWS or HWS) would be used in the actual performance model, which is depicted on the figures as a bold line.

Figure 5.2 shows the dependence of η on the wet-bulb temperature T_W. Over the range tested, both models show a nearly linear dependence on T_W.

Figure 5.3 demonstrates the dependence of η on sprayed temperature T_{HOT}. The HWS model shows a nearly linear dependence, whereas the LWS model has a decreasing slope with increasing temperature.

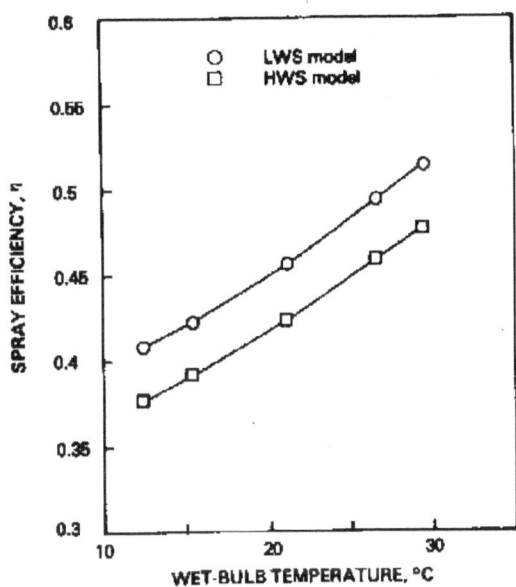

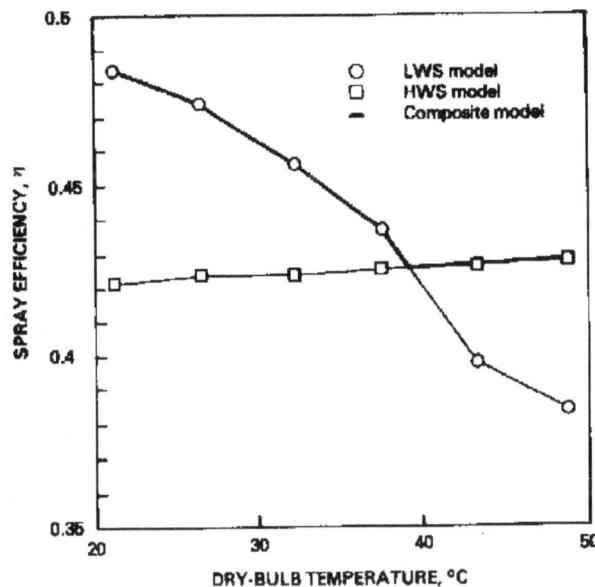

Figure 5.2   Dependence of η on
wet-bulb temperature

Figure 5.3   Dependence of η on
dry-bulb temperature

Figure 5.4 demonstrates the dependence of η on dry-bulb temperature T_A.   The
HWS model shows a small positive, nearly linear dependence on T_A.   The LWS model
shows a much larger, negative dependence with an apparent inflection.

Figure 5.5 demonstrates the dependence of η on windspeed w.   Since windspeed
is not one of the independent variables in the LWS model, η is a constant for
that model.   The HWS model shows a decreasing slope with increasing windspeed.

It is possible to guess a form for the equations (with as-yet-undetermined
coefficients), which would predict the performance of the HWS and LWS models
over a wide range of variations of the independent variables T_A, T_W, T_{HOT},
and w.   The proposed equation for the efficiency of the HWS model would be:

$$\eta_{HWS} = a_1 + b_1 T_A + c_1 T_W + d_1 T_{HOT} + e_1 w + f_1 \sqrt{w} \qquad (5.9)$$

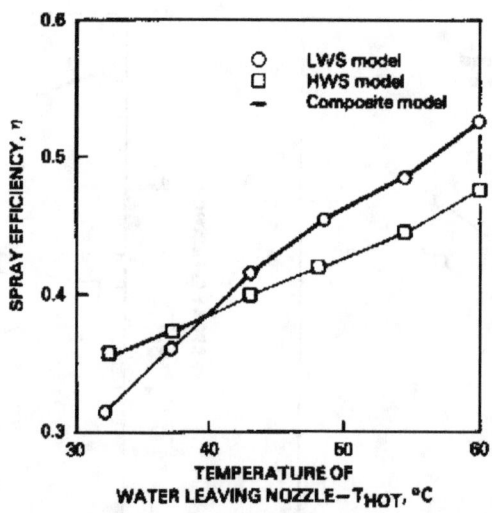

Figure 5.4   Spray efficiency vs sprayed
temperature

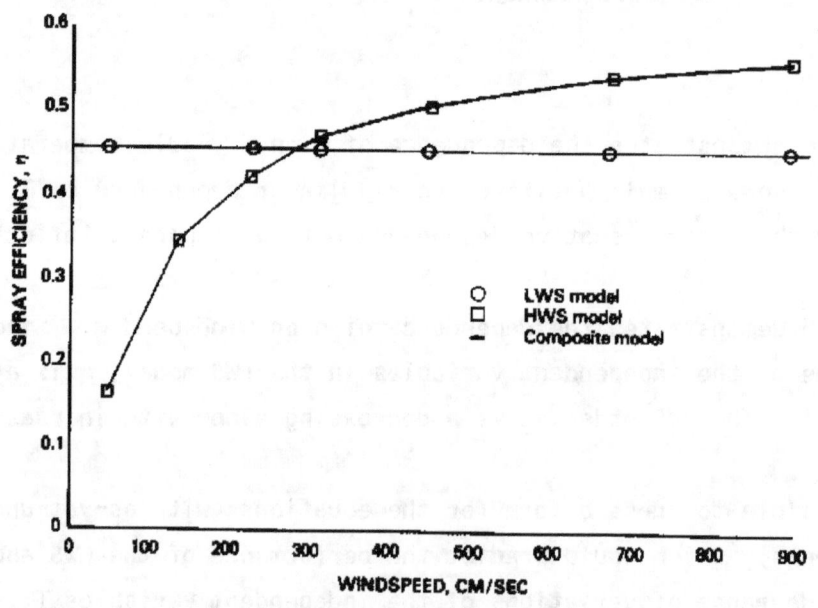

Figure 5.5   Spray efficiency vs windspeed

For the LWS model, the regression equation for efficiency would be:

$$\eta_{LWS} = a_2 + b_2 T_A + c_2 T_A^2 + D_2 T_A^3 + e_2 T_W + f_2 T_{HOT} + g_2 T_{HOT}^2 \qquad (5.10)$$

The evaporation rate Q is correlated in exactly the same fashion:

$$Q_{HWS} = a_3 + b_3 T_A + c_3 T_W + d_3 T_{HOT} + e_3 w + f_3 \sqrt{w} \qquad (5.11)$$

and

$$Q_{LWS} = a_4 + b_4 T_A + c_4 T_A^2 + d_4 T_A^3 + e_4 T_W + f_4 T_{HOT} + g_4 T_{HOT}^2 \qquad (5.12)$$

The coefficients a through g are determined by a least-squares multiple-linear-regression analysis of η and Q over a wide range of the independent variables T_A, T_W, T_{HOT}, and w for the spray pond under investigation. Program SPRCO generates random values of the independent variables in given ranges, runs the HWS and LWS models to generate η and Q, performs the multiple-linear regressions, and presents the correlations of the curve-fitted η and Q versus the calculated η and Q in terms of the coefficient of determination r^2 and a graphical x-y scattergram. The coefficients for Eqs. 5.9 through 5.12 are punched for subsequent use in programs SPSCAN, SPRPND and COMET2. Correlations of the regression equations with the HWS and LWS models are generally excellent.

6 ONSITE-OFFSITE CORRELATION

Long-term meteorological records at the site itself are not usually available
and current NRC practice requires only limited onsite data collection. Furthermore, the meteorological data collected onsite may be incomplete for the purposes
of spray-pond analysis.

The meteorological data for UHS performance must be obtained from offsite weather
stations (such as airports) for which long-term records, including solar radiation or cloud cover, are available. The site meteorology may be significantly
different from that of the offsite station, however, because of such reasons
as orographic features or altitude differences. Thus, it is necessary to
determine if serious discrepancies exist between the two sites. We are only
interested, however, in long-term differences between the meteorology of the
onsite and offsite data, and not the short-term, local variations, such as
thunderstorms.

The assumption is made that we can calculate an "average" pond temperature or
water loss based on monthly (or some other period) averages of the important
meteorological parameters for the onsite and offsite data. By comparing the
monthly average pond temperatures or water loss using the onsite data with the
pond temperature or water loss using the offsite data, we can estimate the bias
that would be introduced by using the offsite data in the temperature calculations. The biases estimated by the above procedure can be used as correction
factors for the water losses and peak temperatures calculated using the long-term
offsite data. Experimentation with the models has shown that the proposed
correction factors reliably account for the differences between the onsite and
offsite data sets and are conservative.

The biases in pond temperature and evaporation can further be related to
differences in each meteorological parameter separately. For example, if the
meteorological parameters of the model are T_A, T_W, H_S, and w:

$$\Delta E \cong \Delta E)_{T_A,w,\dot{H}_{SN}} + \Delta E)_{T_W,w,\dot{H}_{SN}} + \Delta E)_{T_A,T_W,\dot{H}_{SN}} + \Delta E)_{T_A,T_W,w} \qquad (6.1)$$

where

ΔE = overall bias in pond temperature between the two data sets, $^\circ F$

T_W = wet bulb temperature, $^\circ F$

and

$\Delta E)_{T_A,w,\dot{H}_{SN}}$ = bias attributable only to the variation in T_W between the data sets, $^\circ F$

$\Delta E)_{T_W,w,\dot{H}_{SN}}$ = bias attributable only to the variation in T_A between the data sets, $^\circ F$

$\Delta E)_{T_A,T_W,\dot{H}_{SN}}$ = bias attributable only to the variation in w between the data sets, $^\circ F$

$\Delta E)_{T_A,T_W,w}$ = bias attributable only to the variation in $\dot{H}_{SN}$ between the data sets, $^\circ F$

Equation 6.1 is extremely useful because it allows a comparison between onsite and offsite data sets, even if one or more parameters are missing. For example, solar radiation is not usually collected on site. The biases attributable to the other variations can be estimated, bearing in mind that no contribution of the solar radiation difference is included.

A brief computer program, COMET2 (COmpare METeorology), has been written which evaluates the differences in steady-state temperatures between two data sets and their sensitivity to differences in the averages of wet bulb, air temperature, windspeed, and solar radiation between the two sets of data.

This program also calculates the correction factor, in cubic feet of water, for the differences in evaporation and drift between two sites based on the 30-day average meteorology.

Resultant steady-state temperatures and water-loss rates between the two data sets are correlated and the coefficients of correlation, r^2, and the standard error, σ, are calculated.

7 DESCRIPTION OF COMPUTER PROGRAMS

Five separate computer programs are described that are used for several facets
of the spray-cooling-pond analysis:

(1) Program SPRCO simulates the high- and low-windspeed versions of the spray-
     pond-cooling model and generates regression equations based on these models
     for use in subsequent programs.

(2) Program DRIFT calculates a table of drift water loss versus windspeed for
     the spray pond.

(3) Program SPSCAN scans a weather-record tape to predict the likely periods
     of lowest cooling performance and highest evaporation and drift losses.
     Programs DRIFT and SPRCO generate necessary inputs on the pond performance
     for this code.

(4) Program COMET2 compares the limited quantity of onsite meteorological data
     with summaries of offsite data provided by program SPSCAN to determine if
     there are significant differences between the two which might lead to
     differences in predicted pond performance, and suggests correction factors.

(5) Program SPRPND calculates the most pessimistic cooling-pond temperature
     for a design-basis accident using the abbreviated data provided by program
     SPSCAN.

The complicated manner in which these programs are used to determine design-
basis temperature and heat loads is shown in Figure 7.1 and described below.

7.1 Program SPRCO

This program generates the coefficients of a set of multiple-linear regression
equations which represent the cooling performance and evaporative water loss

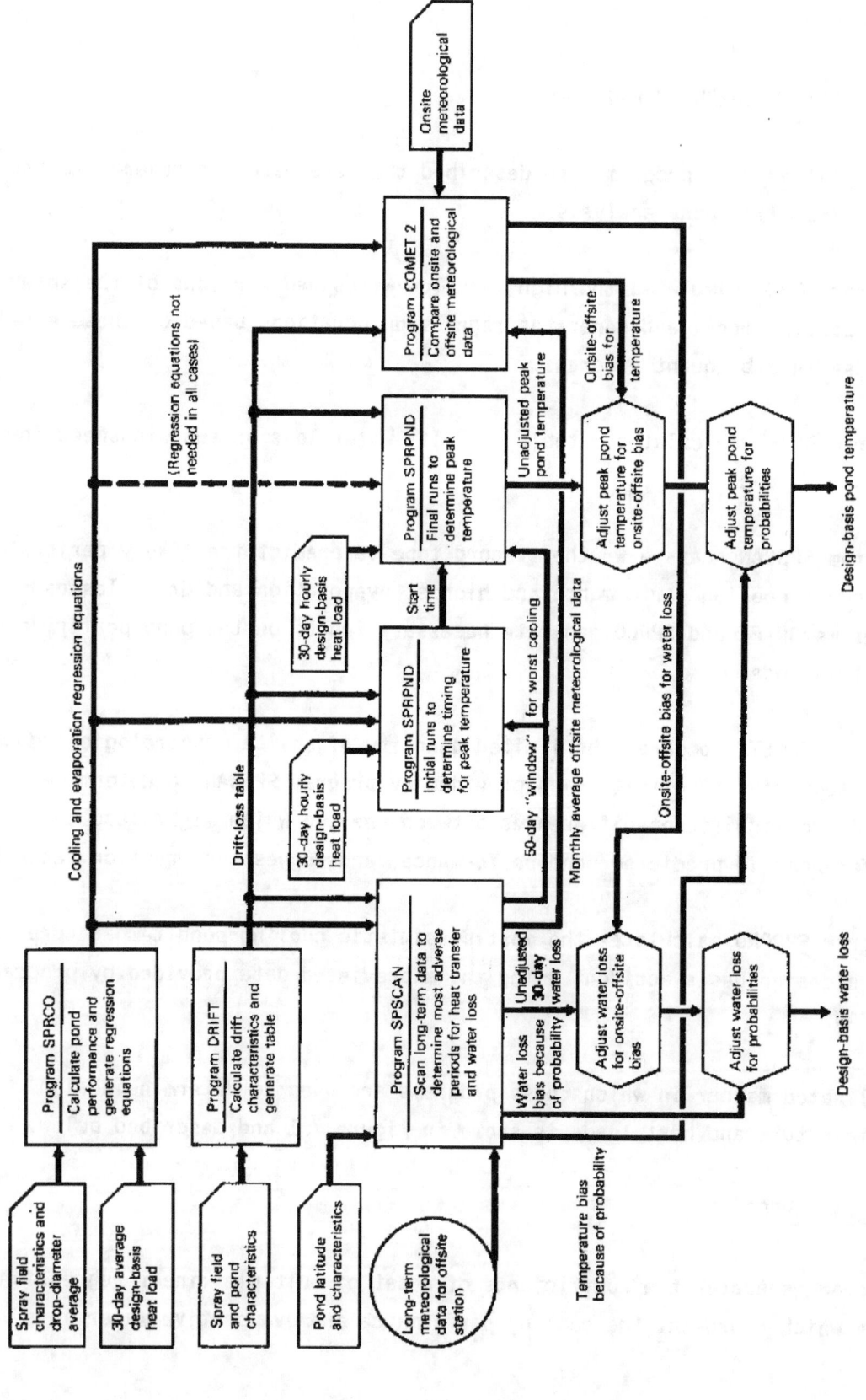

Figure 7.1 Flowchart for design-basis water loss and temperature determination

7-2

of a spray field. The regression equations are subsequently used in programs SPSCAN, SPRPND and COMET2 because they are much less time consuming than the direct use of the HWS and LWS models.

7.1.1 Operation of Program

Program SPRCO runs the HWS and LWS for a large number of cases, typically 200. Each case has the meteorological inputs of wet-bulb temperature T_W, dry-bulb temperature T_A, windspeed w, and sprayed-water temperature T_{HOT}, chosen from a specified range by a pseudorandom-number routine. The resulting cooling performance and evaporation for the HWS and LWS models are recorded and are subsequently fitted to multiple-regression equations whose independent variables are T_A, T_W, w, and T_{HOT}, or powers thereof. Goodness of the fit is tested by calculating the estimated efficiency and evaporation from the four regression equations and comparing these to the results of the HWS and LWS models directly. The standard errors and coefficient of correlation are also calculated.

7.1.2 Program Inputs

Inputs to program SPRCO are of two types:

(1) Variables which describe the basic characteristics of the spray field.

(2) The ranges of meteorological conditions from which each case is randomly chosen.

All inputs are specified in a namelist called INPUT, which is described in Table 7.1. Default values are given where possible, which are typical of Spraco 1751 nozzles with the manufacturer's recommended setup. Only those variables different from the default values need to be read in.

Table 7.1 Namelist INPUT--Inputs to Program SPRCO

Variable name	Description and units	Default value
NPNTS	Number of randomly chosen cases in set	200
VELØ	Initial velocity of drops leaving the spray nozzle, ft/sec	-
HT	Height of spray field from water surface to highest point attained by drop, ft	-
ALEN	Length of the spray field, ft (longer dimension)	-
WID	Width of the spray field, ft (shorter dimension)	-
THETA	Angle of spray to horizontal, degrees	71
YØ	Height of spray nozzles from water surfaces, ft	-
R	Mean drop radius, cm (see text)	0.104
PB	Atmospheric pressure, in. Hg	29.92
Q	Flowrate of water sprayed, ft^3/sec	-
PHI	Heading of wind with respect to long axis, degrees	90
TWETØ	The lower limit of T_W, °F	50
RTW	Range of T_W, °F	30
DTDRYØ	The lower limit of ΔT_A $\left(\text{which is added to the value of } T_W, \text{ since } T_W \geq T_A; \text{ i.e., } T_A = T_W + \Delta T_A\right)$, °F	20
RTD	Range of ΔT_A, °F	30
WINDØ	The lower limit of w, mph	0.1
RW	Range of w, mph	20
THOTØ	The lower limit of T_{HOT}, °F	90
RTH	Range of T_{HOT}, °F	30

Ø = zero

7.1.3 Program Outputs

The following outputs are generated:

(1) The random meteorological inputs and results of the HWS and LWS models for each case.

(2) The regression equations in terms of the coefficient a_1 through g_4:

(a) HWS efficiency (approach to wet bulb):

$$\eta_{HWS} = a_1 + b_1 T_A + C_1 T_A + d_1 T_{HOT} + e_1 w + f_1 \sqrt{w}$$

(b) HWS evaporation (fraction sprayed evaporated):

$$EVAP_{HWS} = a_2 + b_2 T_A + c_2 T_W + d_2 T_{HOT} + e_2 w + f_2 \sqrt{w}$$

(c) LWS efficiency:

$$\eta_{LWS} = a_3 + b_3 T_A + c_3 T_A^2 + d_3 T_A^3 + e_3 T_W + f_3 T_{HOT} + g_3 T_{HOT}^2$$

(d) LWS evaporation:

$$EVAP_{LWS} = a_4 + b_4 T_A + C_4 T_A^2 + d_4 T_A^3 + e_4 T_W + f_4 T_{HOT} + g_4 T_{HOT}^2.$$

(3) Goodness of fit of the regression equations versus the HWS and LWS model outputs:

(a) Coefficient of determination r^2.
(b) Standard error σ.
(c) x-y scattergrams.

7.2 Program DRIFT

The computer program DRIFT computes the drift loss from a spray pond in terms of a fraction of the total amount of water sprayed. The program requires the input of the spray-field geometry and outputs the drift-loss fraction for various windspeeds between 0 and 50 mph. The default drop-diameter distribution in the program is for the Spraco 1751A nozzle under standard operating conditons. Other distributions may be entered.

The spray-field geometry is described by specifying the distances downwind from a group of sprays to the edge of the pond surface and the fraction of the total flow of the spray field represented by that group. When concerned with finding the worst-case drift loss, the direction of the wind is assumed to be the direction that minimizes the distance between the sprays and the edge of the pond surface.

The description of the spray geometry is fairly straightforward when rows of sprays are set parallel to the edge of the pond, as each row can be considered a group of sprays. Irregularly shaped ponds or complex spray arrangements may require an arbitrary grouping of sprays. Figure 3.3 shows how this can be done for a complicated geometry.

To begin the calculation of drift loss from a spray pond, it is first necessary to choose a worst-case wind direction. For simple ponds, this may be done by inspection; more-complex ponds may require that several likely wind directions be modeled before the worst-case wind direction can be determined. Second, the spray field is divided into groups of sprays which are roughly equidistant from the downwind edge of the pond. For conservatism, all of the sprays in the group may be assumed to lie on the boundary of each segment nearest the pond's edge. The fraction of sprays in each group is then calculated.

The final step in the calculations is to prepare the input for DRIFT and run the program. Table 7.2 shows the input format for program DRIFT.

These cards form a repeatable data set. Several runs may be made in a single execution of the program enabling, for example, different pond geometries to be tested.

7.3 Meteorological Data Screening Program SPSCAN

Program SPSCAN is used to scan long-term weather records to determine the period of lowest cooling performance and highest water loss for spray cooling ponds in UHS service. A simple mixed-tank hydraulic model is employed in a running

Table 7.2 Input Variables for Program DRIFT

Card no.	Format	Variables	Comments
1	80A1	TITLE	Columns 2-80 are used to input a message which will be printed at the beginning of the output
2	namelist DROPSZ	DIAM(I), PROPOR(I)	Table for optional drop-diameter distribution. DIAM(I) = drop diameter, cm. PROPOR(I) = corresponding fraction by mass of that diameter. Up to 21 values in table.
3	I2	NUM	Number of cards used in the description of the spray geometry
4 to (3 + NUM)	2F10.0	SPRAY (N,1) SPRAY (N,2)	Distance between a group of sprays and the downwind edge of the pond (ft) and the location of the sprays in the group. There should be NUM cards of this type, one for each group of sprays
(4 + NUM)	80A1	TITLE	The letter "s" is entered in the first column of the last card in the data deck to stop the program

simulation for the entire length of the weather record. Heat and water losses from the sprays are estimated from regression equations generated from program SPRCO and the drift-loss table generated from program DRIFT. The time of maximum ambient pond temperature and the 30-day period giving maximum water loss are determined from the simulation. Annual event statistics are generated for water loss and temperature maxima.

7.3.1 Program Operation

The program first reads and screens meteorological data from National Weather Service Tape Data Family-14 (TDF-14) magnetic tapes. Hourly or three-hourly values of up to 48 meteorological variables are stored on these tapes in a compact alphanumeric code. The program interprets the code and extracts the values of windspeed, dry-bulb temperature, wet-bulb temperature, dewpoint temperature, cloud cover, and atmospheric pressure.

The stored data are checked for missing or inconsistent values. If one or two consecutive observations of a meteorological parameter are missing, they will be replaced by interpolated values. If, however, more than two consecutive observations are missing or in error, the entire day of data is skipped and an informative message to this effect is printed.

The program synthesizes solar radiation needed for subsequent calculations from the cloud cover, date, and latitude, since no direct observations of solar radiation are contained in the TDF-14 tapes. This procedure is discussed in Section 5.2. Direct observations of solar radiation would be most desirable if available from other sources, but no provisions for their input are presently incorporated in the program.

The program then calculates the ambient pond temperature and evaporative loss with the mixed-tank model using the meteorological variables generated in subroutine SUB1. It is necessary to specify a base heat load, which should be the 30-day average design-basis heat load, because the spray performance models are highly nonlinear and sensitive to heat input. The yearly maximum pond temperature and yearly maximum 30-day evaporative and driftwater loss are determined along with their dates of occurrence.

The program statistically treats the data base consisting of the annual maximum pond temperatures and 30-day evaporations. The recurrence interval of the maximum water loss and temperature can be determined from this analysis.

7.3.2 Program Outputs

The program provides the following information, depending in some cases on the options selected:

(1) An informative message is printed if bad data are encountered, so that it is clear that the record for that day has been skipped.

(2) A table of hourly values of windspeed, dry-bulb temperature, wet-bulb temperature, solar radiation, atmospheric pressure, and dewpoint temperature is printed and/or punched (or stored in some other fashion) for the

7-8

20 days preceding the time of maximum ambient temperature and 30 days following. This table may subsequently be used in a more rigorous computation of thermally loaded pond temperature with program SPRPND or some other dynamic temperature model.

(3) The dates and quantity of the yearly worst-30-day-water-loss period for the spray pond with steady heat load is outputted. Since the 30-day-average design-basis heat load is used in this program, the water loss calculated in SPSCAN approximately adequately reflects the design-basis loss (other than seepage) without the need for subsequent modeling.

(4) Monthly averages of meteorological parameters for all specified years of tne record are printed for the purpose of comparing offsite data with limited quantities of onsite data using program COMET2 which will be described later.

(5) The maximum annual pond temperatures and 30-day water losses for all years on the tape are printed, ranked from highest to lowest magnitude. Approximate probabilities are calculated so that the ranked outputs can be plotted on an arithmetic-probability scale. The mean and standard deviation of the data are also printed. Maximum likelihood and confidence limit curves are generated for the statistical adjustment of the design-basis water loss and temperature, as discussed in Appendix A.

7.3.3 Program Inputs

The following input data are necessary to run program SPSCAN:
(1) Pond surface area
(2) Pond volume
(3) Base heat load
(4) Latitude
(5) A TDF-14 weather tape from a representative station near the site

The TDF-14 weather tapes can be obtained for U.S. weather stations from the National Climatic Center, Federal Building, Asheville, North Carolina 28801.

Computer and peripheral requirements to run program SPSCAN on the Brookhaven
National Laboratory CDC 7600 computer are one magnetic tape drive, two disk
files and about 12,000 (decimal) words.

The data deck required to operate program SPSCAN consists of four types of data
cards: the regression coefficient cards, the pond data card, the monthly average
card, and the end card.

The regression coefficients for the spray performance equations are inputted
in exactly the format in which they are punched by program SPRCO. There are
26 variables, read in format 4E15.8 on 7 cards.

The pond data, monthly average, and end cards are read in a namelist format
called INPUT. The variables in this namelist are described in Tables 7.3 and 7.4.

7.3.4 Pond Data Card

This card specifies the pond parameters for the mixed-tank models and specifies
certain printing options as shown in Table 7.3.

Table 7.3 Namelist INPUT--Pond Data Card for Program SPSCAN

Variable name	Value	Type and description
N	1-99	Integer--card number used to identify the card as a "pond data" card and to identify the results in the output
A	≥ 0	Real--pond surface area in square feet
	< 0	In acres
V	≥ 0	Real--pond volume in cubic feet
	< 0	In acre feet
LAT	25-50	Real--latitude of pond in decimal degrees north latitude
IPRNT		Integer--print option
	0	Prints and punches hourly meteorological data
	1	Printed output only
	-1	Punched output only
HEAT		Real--base-heat load, Btu/hr

7.3.5 Monthly Average Card

This card specifies the year and month to start computing monthly meteorological summaries to be used for comparison with onsite meteorological data in program COMET2, as shown in Table 7.4.

Table 7.4 Namelist INPUT--Monthly Average Card for Program SPSCAN

Variable name	Value	Type and description
N	Greater than 99	Integer--identifies this card as a "monthly average" card
YRMODY(1)	-	Real--the year of the beginning date for the computation of monthly averages of meteorological data
YRMODY(2)	5-9	Real--the month of the beginning date for the computation of monthly averages
LAT	25-50	Real--the latitude in decimal degrees north if different from that previously specified

7.3.6 End Card

By specifying $N = 0$, the program terminates.

One set of output is generated from each pond data card or monthly average card. These cards are unrelated and may be inserted in any order.

If a second pond data or monthly average card is used, say, to test the sensitivity to a variation in a pond parameter, only the variable changed needs to be inputted on the namelist card.

7.4 Program COMET2

Program COMET2 (COmpare METeorology) compares steady-state temperature, drift and evaporation rates computed from monthly average values of solar radiation, dry-bulb temperature, wet-bulb temperature, rms windspeed, and barometric pressure for two data sets.

Program SPSCAN computes the monthly averages of the meteorological parameters from the offsite weather station record provided on the National Climatic Center tape. The other data set would be taken from limited onsite measurements.

If onsite data are not complete (for example, if solar radiation is not available), the offsite data can be substituted for the missing parameters. The program calculates the steady-state temperature and 30-day water loss for each data set, the difference in calculated values of pond temperature, and the apparent differences in pond temperature due to differences between each of the meteorological parameters. Therefore, if one of the meteorological parameters for the site is unknown, the apparent differences due to only the other three parameters can still be determined.

The output values of onsite and offsite equilibrium temperature and evaporation rates are correlated for as many months as available to determine if there is a significant difference between the locations. The coefficient of determination r^2 is computed for pond temperatures and water losses for both onsite and offsite locations. A coefficient of determination of 0.9 would indicate that 90% of the variance in one data set is accounted for by variation of the other data set, and that 10% of the variation is unexplained.

The average equilibrium temperature difference and water loss rate difference between the two data sets are the _biases_. The biases may be used cautiously as correction factors to the peak thermally loaded-pond temperature and 30-day evaporation loss. The coefficient of determination r^2 should be high. Lower values may indicate poor quality data or real orographic differences between the sites. Because the data bases are generally small and may be incomplete, it is suggested that the biases be used only in the conservative sense; that is, if onsite values for pond temperatures or water losses are greater than corresponding offsite values, the difference should be added to the peak loaded-pond temperature or water loss as a correction. If the opposite is the case, no corrections should be made.

7.4.1 Program Inputs

Program COMET2 requires recording of monthly averages of dry-bulb temperature, wet-bulb temperature, solar radiation, rms windspeed, and barometric pressure for each site. The first card specifies the number of months of data (I), and is read in I5 format. The next I cards contain the information shown in Table 7.5.

Table 7.5 Meteorological Data Input for Program COMET2

Field	Variable name	Description
1	TW1	Wet-bulb temperature, °F, data set 1
2	TA1	Dry-bulb temperature, °F, data set 1
3	W1	Rms windspeed, mph, data set 1
4	H1	Solar radiation, Btu/(ft^2 day), data set 1
5	PB1	Atmospheric pressure, in. Hg, data set 1
6	TW2	Wet-bulb temperature, °F, data set 2
7	TA2	Dry-bulb temperature, °F, data set 2
8	W2	Rms windspeed, mph, data set 2
9	H2	Solar radiation, Btu/(ft^2 day), data set 2
10	PB2	Atmospheric pressure, in. Hg, data set 2

7.5 Program SPRPND

Program SPRPND calculates the temperature in the UHS pond under the combined influence of the meteorology and the external plant heat load. Hourly meteorological data are provided on cards, disk, or tape from program SPSCAN. The pond is represented by a simplified mixed-tank model used in the screening program SPSCAN. Maximum temperature is determined and the time of the occurrence of the maximum is printed.

7.5.1 Input to Program

Necessary input data for this program include a title card, the external heat input, meteorological conditions, volume and surface area, makeup, blowdown, leakage, circulation flowrate of the pond, height, length, and width of the spray field, and other parameters that describe the sprays.

The first data set consists of the spray performance and evaporation coefficients for the regression equations, punched directly from program SPRCO. There are 26 numbers, read in 4E15.8 format on 7 cards. The spray-pond performance can be calculated from either the regression equations or the self-contained HWS and LWS models, but these seven cards, or seven blank cards, must be read in.

The input data pertaining to the spray field itself are next read in from namelist PARAM, which is defined in Table 7.6.

Table 7.6 Namelist PARAM, Spray-Field Data for Program SPRPND

Parameter	Default value	Description
NDRIFT	-	Number of points in drift-loss table
WDRØ	-	Lowest windspeed in drift-loss table, mph
DWDR	-	Windspeed increment of table, mph
FDRIFT	-	Array of drift-loss fractional values
CEMAX	0.1	Maximum allowed evaporation fraction
CEMIN	0.0	Minimum allowed evaporation fraction
CMAX	0.8	Maximum allowed spray efficiency
CMIN	0.2	Minimum allowed spray efficiency
VELØ	22.5	Initial velocity of drop leaving nozzle, ft/sec*
THETA	71.0	Initial angle with respect to horizon of drop leaving nozzle, degrees*
R	0.104	Average drop radius, cm*
HT	-	Height of spray field, ft*
WID	-	Width of spray field, ft (short dimension)*
ALEN	-	Length of spray field, ft (long dimension)*
YØ	5.0	Height of sprays above water surface*, ft
PHI	80.0	Angle of wind direction with respect to long axis, degrees*
ISPRAY	2	If ISPRAY = 1, use regression model for spray performance
		If ISPRAY = 2, use rigorous model

- = no default value
* = these variables need to be read in only for rigorous model, i.e., ISPRAY=2
Ø = zero

The meteorological data are inputted next. Meteorological data are generally provided directly from program SPSCAN. The first card in the meteorological deck specifies the number of time periods in the table and is read in I5 format. The subsequent cards are read two time periods (usually 1 hr each) per card in the format shown in Table 7.7 as punched by program SPSCAN. (Typically, the meteorological table itself would be stored on a disk or tape file rather than on punched cards. In the present version of the program, this table is read from logical file number 8.)

Table 7.7 Meteorological Input for Program SPRPND
[Format (I3, 3F5.0, F6.0, F7.0, F7.0,
3F5.0, F6.0, F7.0, F7.0)]

Field	Variable	Description
1	ISEQ	Sequence number--not used
2	W(I)	Windspeed, mph
3	TA(I)	Dry-bulb temperature, °F
4	TD(I)	Dewpoint temperature, °F
5	HS(I)	Solar radiation Btu/(ft^2 day)
6	TW(I)	Wet-bulb temperature, °F
7	PRESS(I)	Atmospheric pressure, psia
8	W(I+1)	Windspeed, mph
9	TA(I+1)	Dry-bulb temperature, °F
10	TD(I+1)	Dewpoint temperature, °F
11	HS(I+1)	Solar radiation, Btu/(ft^2 day)
12	TW(I+1)	Wet-bulb temperature, °F
13	PRESS(I+1)	Atmospheric pressure, psia

The heat-and-flowrate table is inputted next. The plant-heat rejection and UHS flowrate during the design accident should be plotted on a log-linear scale, with heat and flowrate on the linear scale and time on the logarithmic scale. A table of heat and flowrate to the pond versus time should then be created from a straight line approximation of the graph. This procedure must be followed because a log-linear interpolation of the heat and flowrate table is used in the program. Also, plant-heat rejection is often provided directly in this graphical form.

7-15

Heat and flowrate are inputted in a namelist format named HFT as shown in Table 7.8.

Table 7.8 Namelist HFT for Program SPRPND

Variable name	Description
HEAT	An array of values of the heat load on the pond, Btu/hr
FLOW	An array of values of the flowrate through the sprays, ft^3/hr
TH	The array of values of time corresponding to the element of the HEAT and FLOW arrays, hr
NH	The number of entries in the table (maximum of 20)

It should be noted that the start of the heat and flowrate table does not necessarily have to correspond to the start of the meteorological input table. The time for the start of the heat-and-flowrate table is delayed by a variable TSKIP(hr) to be described.

Pond parameters and constants are read next in a namelist format called INLIST. The variables in INLIST are described in Table 7.9.

Multiple runs may be made by inserting several title and INLIST cards in succession. Only the variables that are different from the previous namelist card read are changed. A blank title card terminates the program.

7.5.2 Usage of Program SPRPND

Program SPRPND is usually employed to determine maximum pond temperature in the following manner:

(1) Two initial pond simulations should be performed (in the same run):

(a) The first run simulates the pond ambient temperature resulting only from meteorological inputs with a constant base heat load H1 and flowrate F1 specified.

Table 7.9 Namelist INLIST for Program SPRPND

Variable name	Default value	Description
VZERO	0.0	Pond volumes, ft^3--if zero, terminates program
BLOW	0.0	Blowdown flowout, ft^3/hr
A	0.0	Pond surface area, ft^2
NSTEPS	100	Number of timesteps to be performed
NPRINT	10	Printouts of pond temperature and volume every NPRINT steps
DT	0.2	Integration timestep, hr
TZERO	80	Initial pond temperature, °F
TSKIP	0	Time after start of program that corresponds to start of heat-and-flow table. Shifts this table relative to meteorology table which starts at time zero. For time less than TSKIP, evaporation is suppressed so that the pond volume does not decrease
QBASE	0	Bias to be added to all heat in heat-flow table, Btu/hr
FBASE	0	Bias to be added to all flowrate in heat-flow table, ft^3/hr
Q1	0	Heat load for time less than TSKIP, Btu/hr
F1	1	Flow through sprays for time less than TSKIP, ft^3/hr
HEAT ⎫ FLOW ⎬ NH ⎭	Same as specified by previous input in namelist HFT	Heat-flow table if different from that specified by previous input in namelist HFT
ISPRAY	2	If ISPRAY = 1, uses regression equations for spray performance If ISPRAY = 2, uses HWS and LWS performance models directly
IMET	0	If IMET = 0, regular meteorological table used If IMET = 1, constant values TA, TW, W, TD, HS, and PB are used for dry-bulb temperature, wet-bulb temperature, windspeed, dewpoint, solar radiation, and atmospheric pressure as defined in this namelist
TA	90	Constant dry-bulb temperature, °F
TD	60	Constant dewpoint temperature, °F
TW	70	Constant wet-bulb temperature, °F
W	3	Constant windspeed, mph
HS	1500	Constant solar radiation, $Btu/hr/ft^2$
IEVAP	1	If IEVAP = 0, water level in pond remains constant If IEVAP = 1, normal water loss allowed
TSPRON	0	Delay turning on sprays TSPRON hours. Also maintains full pond until sprays are turned on
NITER	0	Repeat run NITER times, incrementing the value of TSKIP and TSPRON by the value DTITER. Used in procedure 2 to determine maximum pond temperature (see paragraph 8.6.2)
DTITER	5	Increment for iterative procedure above, hr

(b) The second simulation determines the peak pond temperature from the effects of external heat input only. This is done by specifying constant values of the meteorological variables.

(2) A second run is prepared so that peak ambient pond temperature determined from the first simulation will roughly coincide with the peak excess temperature caused by plant input alone:

(a) By inspection of the two previous simulations the times of peak temperature for each are chosen.

(b) The approximate time to delay the start of the heat input TSKIP and TSPRON is then defined:

$$\text{TSKIP} = (\text{time of peak ambient temperature}) - (\text{time of peak excess temperature}).$$

(c) The peak pond temperature should occur at approximately the same time as the peak temperature determined for the steady heat load.

Because of nonlinearities in the pond models, the time to the peak temperature may be shifted. An alternative procedure which increments values of the TSKIP and TSPRON for multiple runs may be preferred for determining peak temperature. (See paragraph 8.6.2.) The difference in the final peak temperatures determined will generally be minor.

Either the regression equations (ISPRAY = 1) or the HWS/LWS performance models (ISPRAY = 2) may be used. The latter option has higher accuracy, but the computations are much more time consuming, and may be prohibitive for more than several runs. The regression equations generally give adequate results.

An example run of all programs from start to finish will be covered in the next section.

8 SAMPLE PROBLEM

8.1 Introduction

A complete study of a hypothetical UHS spray pond was undertaken in order to demonstrate the procedure for evaluating the design-basis performance. Details of pond design and meteorology are taken from no plant in particular, but represent eastern U.S. sites and environments. It would be useful to follow the flowchart in Figure 7.1 as an aid in understanding the procedures used.

A plan view of the pond is shown in Figure 8.1. The design-basis heat load is shown in Figure 8.2. Other parameters characterizing the pond are given in Table 8.1.

The spray nozzles are assumed to be of a type similar to the Spraco 1751A, operating at standard pressure and arranged in accordance with the manufacturer's recommendations but with a somewhat different drop-diameter distribution. The drop-diameter distribution for this nozzle is available in only 10 ranges, and given in Table 8.2.

The 28-year tape record (1948-1975) from Harrisburg, Pennsylvania was ordered from the National Climatic Center, Asheville, North Carolina 28801 in TDF-14 format. The spray pond was assumed to be located at the site of the Susquehanna Nuclear Generating Station, although the pond design and heat loads used are not those of this plant, and should not be directly compared. Approximately 15 months of May-October onsite meteorological data were available from the site for a direct side-by-side comparison with the Harrisburg data.

The design-basis evaluation consists of running five programs sequentially as shown in Figure 7.1:

(1) Program SPRCO estimates the regression equations for spray performance for subsequent use in other programs;

(2) Program DRIFT estimates the drift loss for the sprays in the pond configuration as a tabular function of windspeed;

(3) Program SPSCAN scans the TDF-14 meterology tape to determine the periods of most adverse performances and their recurrence intervals;

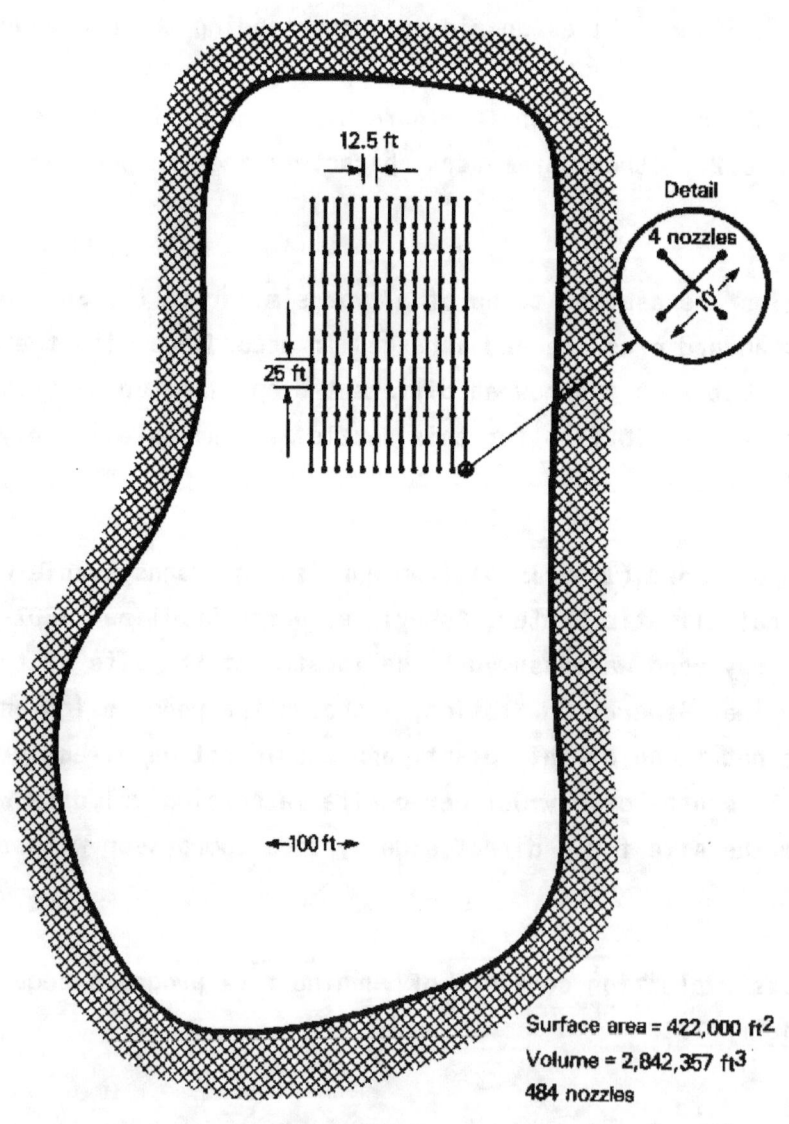

Figure 8.1   Hypothetical spray-cooling pond

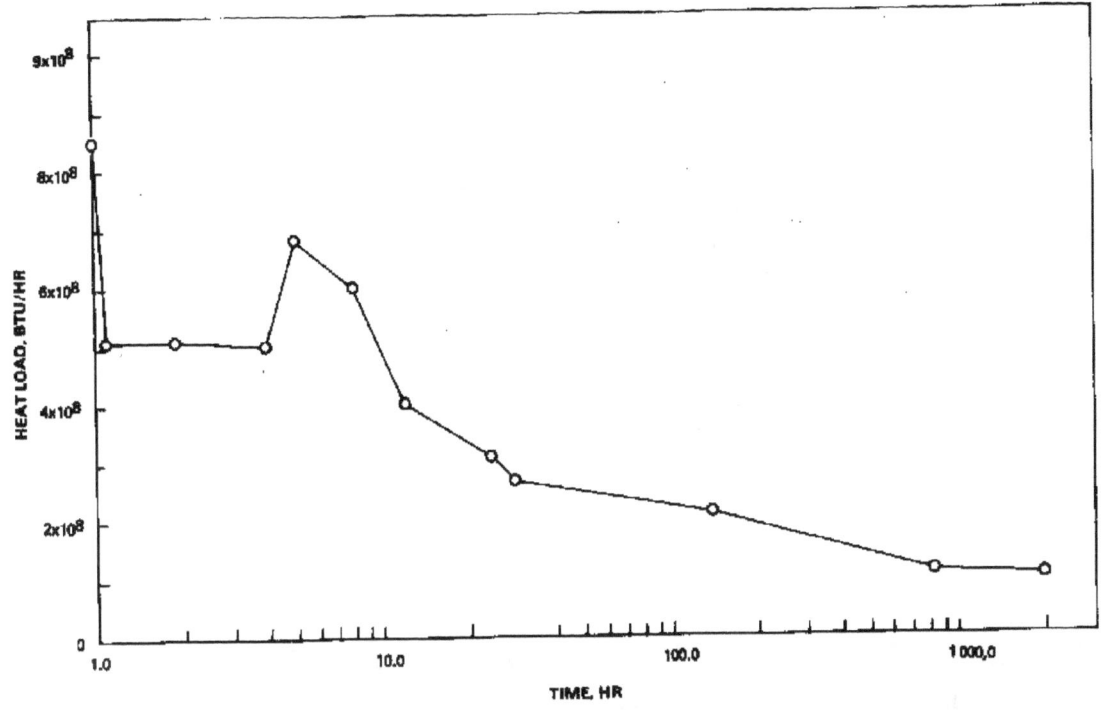

Figure 8.2   Example of design-basis heat load

Table 8.1   Parameters of Spray-Pond Example

Variable	Quantity
Initial pond volume	2,942,357 ft^3
Pond surface area	422,000 ft^2
Flowrate through sprays	57 cfs
Number of nozzles	484
Nozzle pressure	7 psig
Width of spray field	183 ft
Length of spray field	283 ft
Height of nozzles above initial surface	5 ft
Height attained by spray, above nozzles	7 ft

Table 8.2 Drop-Diameter Spectrum
for Spray Nozzle

Diameter, Cm	Volume fraction, %
0.067	10
0.108	10
0.135	10
0.166	10
0.198	10
0.220	10
0.243	10
0.279	10
0.324	10
0.405	10

(4) Program COMET2 compares onsite versus offsite meteorology to predict correction factors for pond temperature and evaporation;

(5) Program SPRPND predicts the uncorrected design-basis pond temperature.

The step-by-step analysis of this spray pond is demonstrated below.

8.2 Determining Characteristics of Spray Field

The first step in the analysis is to determine the inputs for the spray performance model.

8.2.1 Dimensions of Sprayed Region

The average angle of the droplets leaving the spray nozzle can be determined from photographs of sprays operating at the design pressure, or from promotional literature from the spray-nozzle manufacturers. The literature indicated that the spray from the nozzles will reach a height of about 7 ft above the nozzles at a pressure of 7 psig. The heaviest accumulation of water will occur at a radius of about 13 ft. If friction between the drop and the air is neglected, simple ballistics indicates that the initial drop velocity should be about 22.47 ft/sec and the initial angle of the drop trajectory with the horizon should be about 71°.

8.2.2 Average Drop Diameter

The drop-diameter distribution for the nozzle is presented in Table 8.2. Since the distribution is given as tabular values in 10 equal divisions, the discrete summation form of Eq. 2.52 is used for the mean diameter:

$$D_{\frac{1}{2}} = \frac{\sum\limits_{i=1}^{10} \sqrt{D_i}}{\sum\limits_{i=1}^{10} \frac{1}{\sqrt{D_i}}}$$

The mean diameter calculated from the above equation is about 0.19 cm.

8.2.3 Length and Width of Spray Field

The arrangement of sprays is shown in Figure 8.1. The center of each cluster of four nozzles (inset, Figure 8.1) is on 12.5-ft spacing in one direction and 25-ft spacing in the other direction. The overall distance between nozzles is, therefore, 257.1 ft in the long direction and 157.1 ft in the short direction. The actual width of the spray field extends about 13 ft further on each side, which is the radius of the spray umbrella from each nozzle. The length and width of the spray field are, therefore, 283 ft and 183 ft, respectively.

8.3 Spray-Field Performance Regression Equations--Program SPRCO

Program SPRCO generates the coefficients of several regression equations which are used to represent the spray performance models in subsequent programs SPSCAN,

COMET2, and SPRPND. Figure 8.3 shows the input cards for program SPRCO set up in accordance with Section 7.1.

Ranges of meteorological variables were chosen to bound those of the site. Other climates might dictate different ranges. The sprayed temperature chosen is rather high, which places emphasis on the performance of the pond under high-heat-load conditions.

```
$INPUT NPNTS=200,HT=12,ALEN=283,WID=183,VELO=22.47,THETA=71,YO=5,R=.095,
PB=29.92,Q=57,PHI=90,TWETO=50,DTDRYO=20,WINDO=0.1,THOTO=90,RTW=30,RW=20,RTH=30,
RTO=30$
```

Figure 8.3 Input deck for program SPRCO

The output from program SPRCO is shown in Figure 8.4. The high coefficients of determination and relatively small scatter indicate that the regression equations for cooling and evaporative loss should be consistent predictors of the basic performance models. The regression coefficients are outputted on punched cards for subsequent use.

8.4 Determining Drift-Loss Table--Program DRIFT

A table of drift loss versus windspeed is necessary for subsequent use in programs SPSCAN, COMET2, and SPRPND. The arrangement of the spray field with respect to the most critical direction for drift loss is shown in Figure 8.1. Data inputs to program DRIFT are given in Figure 8.5. Although the drop-diameter distribution of Table 8.2 is somewhat different from the Spraco 1751A distribution, only the default distribution is used. Drift is generally only a small contribution to total water loss, and the difference in this case was judged insignificant. If the correct distribution were to be used, a finer division of the scale, especially toward the smaller drop diameters, would be necessary. The output from program DRIFT for the default distribution is shown in Figure 8.6.

COEFFICIENTS FOR EFFICIENCY AND EVAPORATION

INPUT VARIABLES

NUMBER OF RANDOM POINTS,NPNTS = 200
INITIAL VELOCITY OF DROPS LEAVING NOZZLE, VELO = 22.47 FT/SEC
INITIAL ANFLE OF DROPS TO HOR., THETA = 71.000 DEGREES
GEOMETRIC MEAN RADIUS OF DROPS, R = .0950 CM
ATMOSPHERIC PRESSURE, PB = 29.92 INCHES HG
HEIGHT OF SPRAY FIELD, HT = 12.00 FT
WIDTH OF SPRAY FIELD, WID = 183.0 FT
LENGTH OF SPRAY FIELD, ALEN = 283.0 FT
HEIGHT OF SPRAY NOZZLES ABOVE POND SURFACE, YO = 5.0 FT
FLOWRATE OF WATER SPRAYED, Q = 57.00 CU.FT./SEC
HEADING OF WIND W.R.T.LONG AXIS, PHI = 90.00 DEGREES

RANGES OF METEOROLOGICAL PARAMETERS
WET BULB TEMPERATURE = 50.000 TO 80.000 DEG.F
DRY BULB TEMPERATURE = 70.000 TO 130.000 DEG.F
WIND SPEED = .100 TO 20.100 MPH
SPRAYED TEMPERATURE = 90.000 TO 120.000 DEG.F

PT NO.	TWET F	TDRY F	THOT F	WIND MPH	HUMID	ETA LWS	ETA HWS	EVAP. LWS	EVAP. HWS
1	67.4034	115.9188	98.9286	15.8274	.0033	******	.5380	*********	.020377
2	63.6110	83.7989	99.1695	5.6147	.0079	.4149	.3835	.013030	.011946
3	70.6730	102.1529	114.9557	2.7581	.0088	.4544	.3241	.019007	.012913
4	67.4894	90.4482	108.6134	5.6310	.0091	.4553	.4208	.016761	.015325
5	79.3804	120.1969	96.3628	18.7895	.0122	******	.6291	*********	.015147
6	72.9555	121.2013	107.5499	5.8861	.0063	******	.4503	*********	.016422
7	77.0908	123.6912	101.8945	10.0333	.0093	******	.5444	*********	.016790
8	76.1547	124.0878	96.2884	12.0097	.0084	******	.5596	*********	.015672
9	76.6438	110.5680	103.4017	12.6606	.0119	******	.5640	*********	.016678
10	63.7355	90.7825	107.7358	15.3719	.0064	.4359	.5209	.017421	.021192
11	63.3150	98.1664	96.3034	18.5362	.0045	******	.5156	*********	.018110
12	74.9604	99.5406	110.3361	17.0259	.0130	.4561	.5875	.015159	.014981
13	77.7856	108.6317	98.4362	5.1585	.0134	******	.4255	*********	.009660
14	61.3496	84.1540	109.9794	6.3540	.0064	.4549	.4196	.019059	.017558
15	69.5134	104.8221	101.5942	3.7755	.0073	******	.5436	*********	.010932
16	71.3049	112.9713	90.4744	12.9280	.0069	******	.5263	*********	.013812
17	62.6886	94.3131	116.9105	12.1842	.0049	.4635	.5176	.022789	.025712
18	77.2425	114.4053	113.5303	2.9535	.0116	******	.3211	*********	.011077
19	67.6711	96.0390	96.9980	9.1525	.0080	.3888	.4640	.011624	.013746
20	68.7162	112.8469	105.4877	2.8467	.0049	******	.2978	*********	.010872
21	69.0063	111.2167	105.2128	14.1406	.0055	******	.5422	*********	.021321
22	72.4562	109.3174	104.5630	6.2823	.0086	******	.4498	*********	.014952
23	75.6557	121.2607	105.0489	15.8731	.0086	******	.5792	*********	.020161
24	56.0361	81.5676	115.7457	6.0561	.0037	.4605	.4097	.023309	.020749
25	53.6464	80.2143	118.6083	16.3569	.0027	.4632	.5072	.025324	.028258
26	66.0732	103.3842	100.8678	14.7033	.0052	******	.5201	*********	.019210
27	69.3244	104.5041	101.5334	17.8640	.0073	******	.5537	*********	.019146
28	72.9292	119.3568	92.2437	2.6840	.0067	******	.2651	*********	.006386
29	72.9469	120.4440	105.2067	15.3963	.0065	******	.5756	*********	.021750
30	65.6499	114.2207	117.5038	19.2457	.0024	.4336	.5784	.022758	.031003
31	72.5336	102.1859	117.3421	16.1745	.0103	.4762	.5876	.019962	.025103
32	75.0965	105.6050	103.7690	16.1169	.0117	******	.5771	*********	.017556
33	66.2869	112.3171	98.8206	6.9498	.0033	******	.4295	*********	.015601
34	71.0051	92.4416	91.9975	3.8862	.0113	******	.3307	*********	.006774
35	65.8917	100.6496	101.1688	2.2234	.0056	.3897	.2368	.014099	.007846
36	51.0901	76.6584	110.8489	18.4939	.0022	.4355	.4878	.021699	.024855
37	68.6926	117.0634	107.8917	19.2171	.0040	******	.5762	*********	.025237
38	65.7140	95.8163	107.6715	4.8479	.0066	.4219	.3894	.016621	.014989
39	68.8206	101.2450	110.1299	9.0038	.0076	.4174	.4965	.016510	.019756
40	55.7223	93.8110	108.1511	10.5547	.0008	.3989	.4583	.019600	.022675
41	54.4359	89.5847	101.3530	6.2148	.0010	.3737	.3744	.016448	.016277
42	55.4213	83.3247	99.0454	16.4184	.0030	.3893	.4719	.015290	.018918
43	79.4392	99.5306	91.5599	5.9521	.0171	******	.4445	*********	.006209
44	56.5038	92.2766	106.1011	2.1938	.0016	.3966	.2287	.018364	.010011
45	67.7047	110.1276	91.9911	1.0133	.0048	******	.1055	*********	.002781
46	52.2689	84.5510	113.8613	8.9372	.0010	.4367	.4401	.023504	.023800
47	62.1334	104.9040	103.1690	8.3484	.0022	******	.4465	*********	.018832
48	53.5962	90.3111	109.6218	12.3392	.0004	.4116	.4681	.021101	.024257

Figure 8.4   Output from program SPRCO

49	71.8194	97.8476	100.4904	18.8526	.0107	.4079	.5624	.011680	.016444
50	58.3695	91.8028	95.6111	13.6956	.0028	.3502	.4651	.012886	.017369
51	78.6444	102.3428	90.7101	8.5383	.0156	*******	.5035	********	.007661
52	70.3325	91.3590	107.4342	5.5791	.0110	.4555	.4267	.015186	.014074
53	54.9284	92.8665	114.2647	4.2193	.0006	.4291	.3511	.023300	.018558
54	60.1730	90.6156	91.8128	1.4641	.0042	.3554	.1432	.011434	.004165
55	56.1549	88.2818	97.6906	16.4833	.0041	.3950	.4703	.014412	.017541
56	59.4876	87.2057	91.8161	1.1705	.0045	.3226	.1154	.010026	.003352
57	69.0714	113.9063	110.7897	6.2578	.0049	*******	.4510	********	.019149
58	65.6746	113.0471	99.5546	15.6650	.0027	*******	.5278	********	.020772
59	63.5147	101.1542	102.4783	6.5767	.0039	.3891	.4166	.015448	.016132
60	67.0307	88.2855	105.7960	1.7116	.0093	.4471	.2081	.015401	.006861
61	69.6083	109.1265	108.2640	19.0745	.0064	*******	.5753	********	.023480
62	78.2613	125.4515	111.8672	12.6241	.0100	*******	.5945	********	.022694
63	59.9315	106.1264	96.2880	7.9430	.0005	*******	.4171	********	.016579
64	67.5132	100.1693	99.9405	17.1177	.0069	*******	.5360	********	.018191
65	58.9288	95.9033	100.0849	7.3330	.0022	.3621	.4091	.014747	.016489
66	78.1897	113.7561	101.3892	12.5969	.0126	*******	.5717	********	.015552
67	66.6298	94.3767	95.9493	5.7640	.0076	.3723	.3919	.011011	.011268
68	64.2434	113.7548	90.9821	10.5920	.0016	*******	.4648	********	.015929
69	59.8068	93.0472	109.0746	6.1812	.0034	.4207	.4095	.019166	.018429
70	74.3182	96.6462	99.0229	10.4864	.0131	.4120	.5135	.010155	.012674
71	61.0001	105.2545	105.0803	8.0782	.0014	*******	.2795	********	.011768
72	58.0411	82.6889	94.1253	5.4050	.0046	.3732	.3484	.012279	.011297
73	64.3971	110.8943	92.2571	6.5829	.0023	*******	.3999	********	.013257
74	66.9483	96.0381	116.1524	15.1604	.0074	.4713	.5530	.021193	.025213
75	77.3562	102.1617	104.8187	6.5793	.0145	.4364	.4766	.011977	.012820
76	72.9537	113.9999	111.9333	5.9842	.0080	*******	.4618	********	.018170
77	76.9588	111.5965	118.5848	1.5527	.0119	.4717	.2396	.019282	.009035
78	79.0883	100.3738	111.8250	17.4999	.0165	.4787	.6117	.014600	.019093
79	51.1801	83.2446	99.6720	20.0928	.0007	.3790	.4716	.016631	.021198
80	79.9590	100.7945	95.4169	9.8881	.0173	*******	.5322	********	.009141
81	72.7173	122.5141	104.2142	11.6365	.0058	*******	.5434	********	.020278
82	70.1080	100.6791	91.4884	12.1036	.0087	*******	.5025	********	.012442
83	61.4963	86.2209	106.0671	13.9953	.0060	.4347	.4985	.017156	.019997
84	61.9210	97.4075	106.4095	19.1029	.0037	.3876	.5310	.016492	.023274
85	63.8127	100.0225	103.3220	3.4352	.0044	.3901	.3142	.015424	.011732
86	72.0654	121.6405	114.1130	13.4156	.0058	*******	.5744	********	.025972
87	52.4328	86.9093	117.0768	18.4850	.0006	.4441	.5105	.025293	.029624
88	58.0944	100.7889	104.3526	4.5358	.0006	.3755	.3466	.017803	.015413
89	68.6256	90.5076	113.9251	11.0905	.0099	.4833	.5234	.019258	.021075
90	63.5293	105.6638	91.2158	10.2672	.0029	*******	.4545	********	.014903
91	57.9937	85.4488	95.7425	.4734	.0040	.3703	.0473	.012977	.001380
92	57.9713	96.7101	118.1942	9.9329	.0015	.4457	.4832	.024748	.026921
93	65.2281	97.4012	99.3172	19.5465	.0059	.3896	.5329	.013442	.018720
94	72.7199	104.0671	106.0796	6.0870	.0100	.4297	.4485	.014514	.014718
95	68.5792	113.9043	98.6573	6.9392	.0046	*******	.4391	********	.015092
96	63.4203	98.3815	95.7829	13.8597	.0045	*******	.4906	********	.016865
97	56.7413	92.3483	112.4770	19.1774	.0017	.4289	.5211	.021870	.027121
98	79.9125	120.6913	98.2020	4.4492	.0126	*******	.4083	********	.009232
99	78.8285	127.2700	93.1343	11.1476	.0101	*******	.6387	********	.015174
100	64.8418	92.5447	92.9948	10.5771	.0068	.3692	.4599	.010652	.013219
101	64.7569	95.1750	90.5651	19.7177	.0061	*******	.5148	********	.014690
102	70.3287	111.6158	108.0073	4.7942	.0064	*******	.4049	********	.015559
103	79.8324	122.5774	101.1812	13.7402	.0121	*******	.5999	********	.016546
104	50.2843	84.3593	111.8900	4.1588	.0000	.4237	.3312	.022954	.017607
105	67.6379	105.7145	104.8616	4.3632	.0057	*******	.3714	********	.013595
106	76.5129	97.4769	93.7958	3.8063	.0148	*******	.3503	********	.006151
107	50.2871	75.6865	100.6088	13.8789	.0020	.3996	.4415	.017028	.019177
108	61.1566	94.0389	94.6069	13.5144	.0040	.3622	.4740	.012411	.016340
109	61.6523	92.7169	100.8256	19.4719	.0046	.3774	.5171	.014091	.019873
110	68.7779	96.4446	90.5561	10.0251	.0087	*******	.4686	********	.011255
111	69.2385	89.9574	114.4638	13.8699	.0105	.4891	.5460	.019320	.021985
112	79.2581	128.2571	116.6875	2.8362	.0102	*******	.3623	********	.014153
113	51.0051	79.0942	99.0560	.6522	.0016	.3881	.0671	.016336	.002695
114	61.0026	87.8337	113.3350	17.5152	.0053	.4585	.5307	.021103	.024898
115	71.1008	105.0568	104.6273	12.2713	.0085	*******	.5325	********	.018511
116	72.2442	93.8480	112.2074	15.0614	.0120	.4800	.5651	.017231	.020637
117	53.6169	79.2727	113.8384	11.5373	.0029	.4496	.4683	.022772	.024064
118	58.6711	83.3407	103.7567	7.1649	.0049	.4223	.4115	.016708	.016269
119	64.5140	105.3610	93.9220	2.3925	.0036	*******	.2284	********	.006941
120	57.2807	91.5981	96.8422	16.0434	.0022	.3258	.4767	.012430	.018808
121	64.1858	93.1116	103.4380	3.8820	.0062	.4029	.4521	.014877	.016711
122	79.2938	107.4412	109.2870	3.8628	.0151	.4594	.4012	.013957	.011532
123	66.6769	116.0519	100.5806	17.3038	.0027	*******	.5451	********	.021838
124	69.1430	103.5153	113.7618	.5781	.0074	.4301	.0842	.018342	.003336
125	56.3752	84.8528	99.7334	18.1018	.0032	.3690	.4846	.015302	.019508
126	61.9749	88.3719	107.5155	10.3578	.0058	.4367	.4750	.017814	.019569
127	79.0554	103.2812	99.1784	11.1396	.0158	*******	.5493	********	.011949
128	68.7923	98.1392	116.8077	15.5632	.0083	.4741	.5653	.020929	.025380
129	73.2131	108.5415	113.5956	17.6234	.0094	.4494	.5946	.018027	.024338
130	55.4639	82.9496	99.7177	14.8472	.0031	.3939	.4656	.015593	.018775

Figure 8.4 (Continued)

131	61.3000	68.8576	114.7576	11.8224	.0053	.4627	.5029	.021826	.023983
132	74.4032	114.3866	105.0155	4.1364	.0091	*******	.3858	********	.012270
133	51.5690	84.2426	109.8430	13.8297	.0007	.4206	.4693	.021596	.024458
134	50.5931	70.6975	116.9882	11.2832	.0032	.4588	.4612	.024244	.024867
135	52.2620	78.0522	116.2913	11.9360	.0025	.4544	.4720	.024269	.025602
136	61.6710	83.1307	99.7283	19.2615	.0068	.4126	.5079	.013825	.017451
137	73.8796	115.5426	101.2148	13.5128	.0088	*******	.5563	********	.017638
138	51.1270	72.2501	116.4497	17.3699	.0032	.4580	.4947	.024042	.026575
139	72.7124	107.6576	101.4917	5.8419	.0092	*******	.4323	********	.013030
140	78.6340	115.3275	93.9846	16.2058	.0126	*******	.6036	********	.013078
141	56.7777	95.2767	94.6560	15.5823	.0011	*******	.4699	********	.018627
142	64.6105	109.8044	119.4114	13.4093	.0027	.4098	.5453	.021677	.029597
143	71.3983	95.3890	110.7226	18.1067	.0110	.4620	.5743	.016628	.021133
144	77.9664	126.3154	100.1681	9.9117	.0095	*******	.5477	********	.015976
145	78.5490	99.8391	105.1210	8.7062	.0161	.4142	.5223	.010540	.013434
146	57.5637	89.1638	116.5244	11.9659	.0029	.4553	.4944	.023809	.026122
147	73.2727	121.2378	109.8102	3.1316	.0066	*******	.3396	********	.012620
148	64.6853	93.5201	105.9272	15.9243	.0064	.4181	.5253	.016070	.020609
149	68.2226	94.7813	99.4242	8.8357	.0087	.3580	.4662	.010750	.014201
150	74.6831	109.0938	107.2351	1.6844	.0105	*******	.2199	********	.006741
151	66.0542	89.9826	97.0707	15.1379	.0082	.3784	.5066	.011097	.019212
152	74.1215	117.6612	105.9791	7.8658	.0081	*******	.4977	********	.017477
153	76.7173	123.7331	109.4219	18.9267	.0089	*******	.6224	********	.023748
154	57.0949	90.9050	106.2536	10.2591	.0022	.4047	.4551	.018393	.020819
155	58.1275	93.6501	108.8546	13.7410	.0022	.4115	.4935	.019437	.023661
156	54.3402	87.3218	113.2785	7.4716	.0015	.4362	.4255	.022818	.022219
157	71.7842	106.3771	97.3862	3.8282	.0087	*******	.3435	********	.009226
158	75.5373	122.6951	110.5109	2.5525	.0081	*******	.3079	********	.010840
159	57.2252	98.2442	106.6623	15.8426	.0007	.3685	.4992	.017543	.024399
160	56.5473	77.6957	92.6084	8.9358	.0049	.3786	.4036	.011930	.012831
161	64.4577	111.0088	110.1480	7.9576	.0023	*******	.4659	********	.021685
162	56.7155	99.4639	117.3073	13.4489	.0001	.4283	.5069	.024351	.029193
163	72.0759	114.8235	107.5444	16.5400	.0071	*******	.5777	********	.022581
164	55.6898	90.2117	109.4029	19.4244	.0016	.4179	.5103	.020478	.025549
165	62.3659	83.1844	99.3503	13.3041	.0072	.4132	.4612	.013454	.015951
166	62.9510	94.7213	99.7866	17.8437	.0050	.3612	.5154	.012931	.019018
167	66.6190	93.2614	106.3580	4.1365	.0078	.4287	.3628	.015763	.012969
168	79.4957	114.5755	113.9865	11.4705	.0136	*******	.5563	********	.020950
169	68.7722	114.9594	93.8845	5.8484	.0045	*******	.4023	********	.012248
170	64.3319	111.0137	99.7700	12.3209	.0022	*******	.4969	********	.019850
171	79.3533	120.9988	119.5969	14.4833	.0120	*******	.6239	********	.026386
172	56.9720	79.0464	93.8202	7.5818	.0040	.3814	.3890	.012376	.012655
173	58.3138	83.2309	110.1849	10.7160	.0047	.4466	.4704	.019965	.021265
174	51.4455	81.6802	117.7597	2.7979	.0012	.4513	.2883	.025553	.016039
175	58.3238	89.4820	101.8941	8.8074	.0033	.3899	.4324	.015794	.017560
176	75.5881	108.8309	97.5416	13.8105	.0114	*******	.5593	********	.014463
177	63.6032	105.1123	119.6778	12.2361	.0031	.4419	.5313	.023583	.028708
178	65.6732	105.6303	113.1837	8.8961	.0044	.4075	.3138	.018837	.013769
179	65.6493	112.4201	115.5197	13.7300	.0028	.4289	.5450	.021643	.027777
180	58.2750	95.1571	107.5181	18.3230	.0020	.3975	.5146	.018483	.024504
181	60.1580	85.6616	118.8807	1.4109	.0053	.4788	.1974	.024169	.009711
182	76.9950	126.6796	116.0926	12.2467	.0085	*******	.5920	********	.025492
183	69.9647	90.4464	116.5879	10.7991	.0110	.5063	.5361	.021362	.022856
184	57.5646	92.8684	112.1333	11.1285	.0021	.4287	.4788	.021467	.024164
185	72.1245	108.8369	101.4905	17.7975	.0085	*******	.5705	********	.018733
186	61.5944	85.8452	95.7620	18.2366	.0061	.3805	.4975	.012027	.016129
187	63.1549	93.7709	119.5039	15.5678	.0053	.4767	.5461	.024117	.028055
188	65.6625	101.4808	116.4901	17.9258	.0053	.4479	.5645	.021445	.027625
189	72.8020	109.5644	118.9666	13.6729	.0088	.4428	.5808	.019693	.026450
190	57.1963	78.7206	110.3588	10.2728	.0054	.4534	.4634	.019846	.020570
191	65.9348	93.1745	114.5765	17.8526	.0074	.4688	.5569	.020537	.024871
192	72.4141	105.4093	93.4107	11.3346	.0095	*******	.5121	********	.012790
193	70.1000	99.7839	102.0029	5.7930	.0089	.4015	.4203	.012879	.013113
194	50.9828	75.4225	105.9585	13.8405	.0024	.4210	.4556	.019275	.021263
195	69.4827	118.2645	90.2725	4.7112	.0043	*******	.3598	********	.009863
196	71.4134	101.5811	110.0974	16.1022	.0096	.4270	.5680	.015822	.021580
197	73.5714	104.0047	94.4088	5.3200	.0108	*******	.4040	********	.009282
198	64.4607	111.3646	108.5826	7.7821	.0023	*******	.4595	********	.020863
199	71.0673	102.9814	106.7075	14.1714	.0090	.4231	.5495	.015025	.019772
200	59.9966	108.3595	118.1041	11.9669	.0001	.3947	.5128	.022150	.029395

NUMBER OF POINTS GENERATED = 200
NUMBER OF POINTS PLOTTED = 117

FOR HWS EFFICIENCY,CONSTANT AND COEFF OF T,TWET,THOT,
WIND AND WIND**,S ARE
-.60657276E+00
.40195127E+03
.38449863E-02
.18250238E-02
-.34078270E+01
.30138737E+00

Figure 8.4 (Continued)

FOR HWS EVAPORATION,CONSTANT AND COEFICIENT OF T,   TWET,THOT,WIND AND WIND**.5 ARE
 -.41450389E-01
  .14646531E-03
 -.33234415E-03
  .41560445E-03
 -.12268707E-02
  .11416664E-01

FOR LWS EFFICIENCY,CONSTANT AND COEFF OF T,T**2,   T**3,TWET,THOT AND THOT**2 ARE
 -.25690451E+01
  .65576685E-01
 -.73791051E-03
  .26319278E-05
  .35669730E-02
  .12911864E-01
 -.39275022E-04

FOR LWS EVAPORATION,CONSTANT AND COEFF OF T,T**2,   T**3,TWET,THOT AND THOT**2 ARE
 -.86122112E-01
  .28767122E-02
 -.29725976E-04
  .10168749E-06
 -.27394599E-03
  .28406611E-04
  .22034012E-05

PLOTTED CHARACTERS ARE NUMBER OF POINTS FALLING AT THAT POSITION

Figure 8.4   (Continued)

8-10

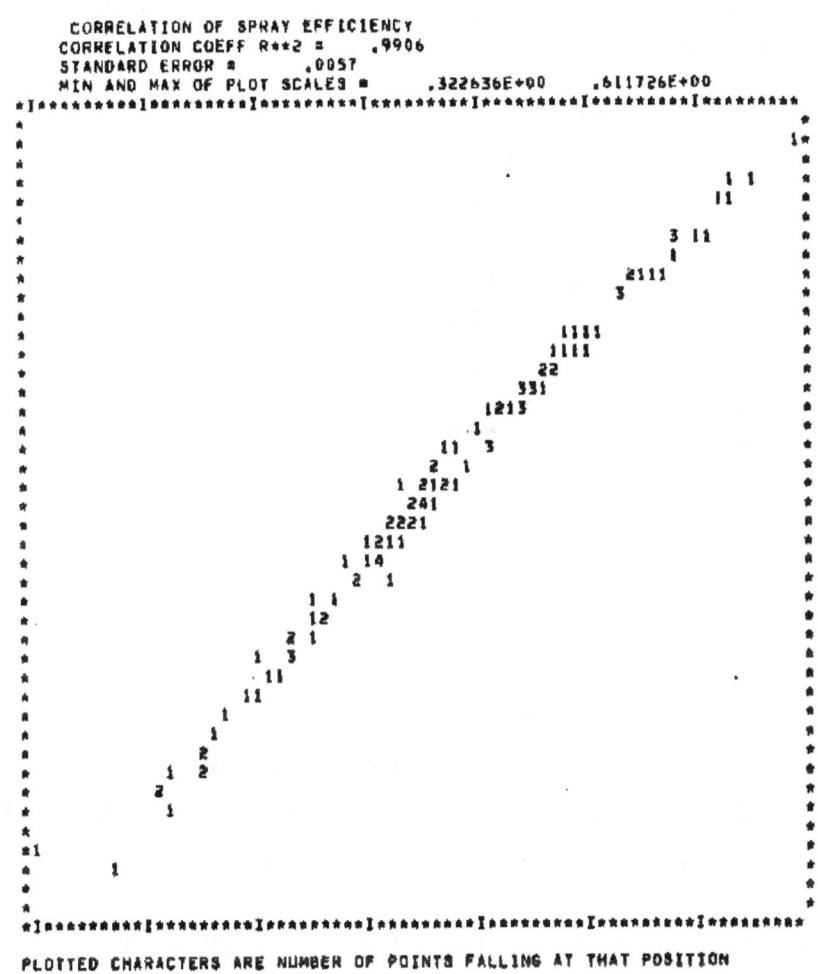

Figure 8.4 (Continued)

8.5 Scanning Weather Record--Program SPSCAN

The periods of most-adverse meteorology with respect to cooling performance
and water loss were determined from the tape meteorological record using program
SPSCAN and the output from programs SPRCO and DRIFT. The inputs to program
SPSCAN were developed according to Section 7.4, and are shown in Figure 8.7.
There is one pond data card, one monthly average card, and one end card.

```
DRIFT TABLE FOR HYPOTHETICAL SPRAY POND
$DROPSZ$
13
    120.0  .07692308
    132.5  .07692308
    145.0  .07692308
    157.5  .07692308
    170.0  .07692308
    182.5  .07692308
    195.0  .07692308
    207.5  .07692308
    220.0  .07692308
    232.5  .07692308
    245.0  .07692308
    257.5  .07692308
    270.0  .07692308
 $
```

Figure 8.5. Input deck for program DRIFT

The base heat load HEAT on the pond data card is taken to be the 30-day average excess heat load from the plant, so that the 30-day evaporative loss calculated in this program could be used directly, since evaporation is approximately proportional to the cumulative heat load.

Partial printed output is shown in Figure 8.8. In addition to the printed output, the hourly record of the 20-day period before and the 30-day period after the time of most-adverse cooling performance was either punched or (preferably) stored on a permanent file for further use in program SPRPND. This output is also shown in Figure 8.8.

8.6 Determining the Uncorrected Design-Basis Temperature--Program SPRPND

Once the period of most-adverse meteorology for cooling has been determined by program SPSCAN, program SPRPND is run to simulate the pond temperature under tne actual design-basis heat loads. One of two procedures may be followed to make this determination.

TITLE: DRIFT TABLE FOR HYPOTHETICAL SPRAY POND

SPRAY GEOMETRY (13 POINTS)

FEET FROM EDGE	FRAC. OF SPRAYS
120.000000	.076923
132.500000	.076923
145.000000	.076923
157.500000	.076923
170.000000	.076923
182.500000	.076923
195.000000	.076923
207.500000	.076923
220.000000	.076923
232.500000	.076923
245.000000	.076923
257.500000	.076923
270.000000	.076923

DRIFT LOSS FRACTION

WIND SPEED	LOSS FRAC.
0.000	.00050000
2.500	.00050000
5.000	.00050000
7.500	.00050000
10.000	.00058047
12.500	.00075946
15.000	.00106712
17.500	.00145037
20.000	.00191420
22.500	.00237861
25.000	.00296085
30.000	.00434594
35.000	.00590310
40.000	.00789034
45.000	.01086714
50.000	.01432954

Figure 8.6. Output from program DRIFT

```
-.60637276E+00    .40195127E-03    .38449863E-02    .18230236E-02
-.34078270E-01    .30138737E+00   -.25690451E+01    .65576685E-01
-.73791051E-03    .26319278E-05    .35669730E-02    .12911864E-01
-.39275022E-04   -.41450389E-01    .14646531E-03   -.33234415E-03
 .41560445E-03   -.12268707E-02    .11416664E-01   -.86122112E-01
 .28767122E-02   -.29725976E-04    .10168749E-06   -.27394599E-03
 .28406611E-04    .22034012E-05
$INPUT N=1,A=422000.,V=2942357.,LAT=41.2,HEAT=2.3E8,IPRNT=0,WDRO=0,
   NDRIFT =6,QWDR=10,FDRIFT=.0005,.00058,.001914,.004346,.007890,.014330$
$INPUT N=100,YRMODY(1)=73,YRMODY(2)=5$
$INPUT N=0$
```

Figure 8.7   Input deck for program SPSCAN

8.6.1 Procedure 1

(1) Make two runs of program SPRPND; first to determine the pond temperature
     for the ambient meteorology (but with a steady heat load), and second, to
     determine the pond response to the design-basis heat load, but with constant
     meteorological parameters.

(2) Make a third run combining the time-varying meteorology and heat load,
     with the timing adjusted so that the two temperature peaks determined in
     the step above are approximately superimposed.

The inputs to program SPRPND for the first run are shown in Figure 8.9. The
parameter IMET = 0 specifies that the tabular meteorological data is used for
meteorology.   ISPRAY = 1 specifies that the regression model is used for
spray-heat and mass transfer.   TSKIP = 5000 effectively eliminates the use of
the design-basis heat-and-flowrate table.   Parameter Q1 = 0.23E9 specifies the
steady heat load for this run.   IEVAP = 0 forces the pond to remain full for
the run.   TSPRON = 0 specifies that the sprays are turned on at the beginning
of the run.

The output of run 1 is shown printed in Figure 8.10 and plotted in Figure 8.11.

8-14

********** SUBROUTINE SUB1 HAS BEEN CALLED FOR LATITUDE = 41.20 DEG. NORTH *****

```
DISCONTINUITY IN DATA CAUSED  6/11/71 TO BE SKIPPED
DISCONTINUITY IN DATA CAUSED  9/25/71 TO BE SKIPPED
DISCONTINUITY IN DATA CAUSED  5/ 5/72 TO BE SKIPPED
DISCONTINUITY IN DATA CAUSED  5/ 6/72 TO BE SKIPPED
DISCONTINUITY IN DATA CAUSED  5/ 7/72 TO BE SKIPPED
DISCONTINUITY IN DATA CAUSED  5/ 8/72 TO BE SKIPPED
DISCONTINUITY IN DATA CAUSED  8/ 7/72 TO BE SKIPPED
DISCONTINUITY IN DATA CAUSED  7/11/73 TO BE SKIPPED
DISCONTINUITY IN DATA CAUSED  7/15/73 TO BE SKIPPED
DISCONTINUITY IN DATA CAUSED  7/16/73 TO BE SKIPPED
DISCONTINUITY IN DATA CAUSED  7/17/73 TO BE SKIPPED
DISCONTINUITY IN DATA CAUSED  5/ 1/75 TO BE SKIPPED
```

*********** POND NUMBER 1 HAS THE FOLLOWING PARAMETERS *************************

SURFACE AREA   422000.00 FT**2 (9.69 ACRES)

VOLUME   2942357.00 FT**3 (67.55 ACRE-FT)

ISRCH = 1   IPRNT = 0

********** POND NUMBER 1 HAS BEEN MODELLED TO DETERMINE THE WORST **************
          PERIODS FOR COOLING AND EVAPORATIVE WATER LOSS

***********SPRAY PARAMETERS
          BASE HEAT LOAD =   .23E+09 BTU/HR
          MINIMUM EVAPORATIVE LOSS FRACTION =   0.000000
          MAXIMUM EVAPORATIVE LOSS FRACTION =   .050000
          MINIMUM SPRAY EFFICIENCY =   .1000
          MAXIMUM SPRAY EFFICIENCY =   .8000

***********DRIFT LOSS TABLE

WIND SPEED - MPH	DRIFT LOSS FRACTION
0.00	.000500
10.00	.000580
20.00	.001914
30.00	.004346
40.00	.007840
50.00	.014330

Figure 8.8   Output from program SPSCAN

************THE SAMPLE OF YEARLY MAXIMUM POND TEMPERATURES AND 30 DAY *********
EVAPORATIVE LOSSES GENERATED BY THIS MODEL IS DESCRIBED BELOW.**

..........TEMPERATURE................			............EVAPORATIVE LOSS..........		
*EXCEEDED */100 YR*	(DEG.F)	DATE *(YR.MO.DY.)	*EXCEEDED */100 YR*	FT**3	DATE *(YR.MO.DY.)*
2.45	92.74	72. 7.22.	2.45	2462822.8	66. 7.20.
5.97	92.02	75. 8. 2.	5.97	2409377.5	55. 8. 9.
9.49	91.55	59. 6.30.	9.49	2352520.5	63. 7.30.
13.01	91.11	68. 7.18.	13.01	2337086.9	74. 7.30.
16.54	91.04	73. 8.31.	16.54	2332290.8	57. 7.15.
20.06	91.00	48. 8.27.	20.06	2330110.7	52. 7.24.
23.58	90.98	57. 6.18.	23.58	2329511.1	71. 7.20.
27.10	90.48	52. 7.22.	27.10	2326817.8	68. 8.11.
30.63	90.39	65. 8.17.	30.63	2323800.6	65. 7.20.
34.15	90.35	49. 7.29.	34.15	2309946.7	64. 7. 8.
37.67	90.34	53. 9. 2.	37.67	2306465.0	54. 8.12.
41.19	90.01	62. 7. 8.	41.19	2302173.2	53. 7.20.
44.72	90.01	66. 8.22.	44.72	2302154.9	49. 7. 5.
48.24	89.97	63. 7. 1.	48.24	2288803.4	62. 7.27.
51.76	89.57	55. 7. 4.	51.76	2287736.5	59. 7. 5.
55.28	89.54	64. 7.20.	55.28	2286469.7	72. 7.26.
58.81	89.39	70. 8. 1.	58.81	2283665.6	73. 8. 6.
62.33	89.27	69. 6.28.	62.33	2270155.6	51. 8. 2.
65.85	89.23	60. 8.29.	65.85	2268953.0	75. 8.15.
69.37	89.07	61. 7.23.	69.37	2267994.3	67. 7. 4.
72.90	88.73	71. 6.28.	72.90	2265241.7	70. 8.31.
76.42	88.65	74. 7. 4.	76.42	2261924.0	69. 7.24.
79.94	88.10	67. 6.17.	79.94	2251452.1	56. 7.10.
83.46	88.09	51. 8.10.	83.46	2245862.8	48. 7.24.
86.99	87.95	58. 7.27.	86.99	2242686.2	61. 7.28.
90.51	87.79	56. 8.31.	90.51	2226405.1	60. 7.24.
94.03	87.62	50. 8. 1.	94.03	2215186.3	58. 7.30.
97.55	87.49	54. 9. 6.	97.55	2202979.8	50. 7.21.

MEAN	89.73	2296092.7
STANDARD DEV.	1.380	55542.68
SKEW	.154	1.010

Figure 8.8   (Continued)

PREDICTED VALUES AND CONFIDENCE LIMITS ON
PEAK TEMPERATURE, DEG.F

EXCEEDED PER 100 YR	PREDICTED VALUE	5 PERCENT CONFIDENCE	95 PERCENT CONFIDENCE
.100	94.451	93.288	95.613
.500	93.554	92.477	94.530
1.000	93.142	92.248	94.037
2.000	92.708	91.898	93.518
5.000	92.081	91.386	92.776
10.000	91.544	90.937	92.150
20.000	90.911	90.392	91.430
30.000	90.463	89.949	90.938
40.000	90.084	89.633	90.535
60.000	89.378	88.927	89.830
70.000	88.999	88.525	89.473
80.000	88.551	88.033	89.070
90.000	87.919	87.313	88.525
95.000	87.341	86.606	88.077
98.000	86.754	85.944	87.564
99.000	86.320	85.425	87.214
99.500	85.909	84.932	86.886
99.900	85.012	83.849	86.174

PREDICTED VALUES AND CONFIDENCE LIMITS ON
30 DAY EVAPORATION, FT**3

EXCEEDED PER 100 YR	PREDICTED VALUE	5 PERCENT CONFIDENCE	95 PERCENT CONFIDENCE
.100	2486106.070	2439307.091	2532905.048
.500	2449983.821	2410657.411	2489310.231
1.000	2433430.819	2397419.438	2469442.200
2.000	2415944.033	2383327.751	2448560.315
5.000	2390697.844	2362709.866	2418645.863
10.000	2369059.240	2344656.889	2393461.591
20.000	2343589.274	2322696.192	2364482.356
30.000	2325566.328	2306470.449	2344662.207
40.000	2310303.682	2292134.736	2328472.628
60.000	2281881.643	2263712.697	2300050.589
70.000	2266618.996	2247523.118	2285714.875
80.000	2248596.051	2227702.969	2269489.132
90.000	2223126.085	2198723.734	2247528.435
95.000	2201487.460	2173499.462	2229475.459
98.000	2176241.292	2143625.010	2208857.573
99.000	2158754.506	2122743.125	2194765.887
99.500	2142201.504	2102875.094	2181527.914
99.900	2106079.255	2059280.276	2152878.233

Figure 8.8   (Continued)

********** METEOROLOGY FOR   7/ 3/72***

HOUR	WIND SP. (MPH)	DRY BULB (DEG.F)	DEWPOINT (DEG.F)	SOLAR RAD BTU/FT2/D	WET BULB (DEG.F)	ATM.PRESS. PSIA
0.	5.0	74.0	65.7	0.0	68.67	29.59
1.	3.5	72.0	66.0	0.0	68.00	29.58
2.	4.2	72.0	66.0	0.0	68.00	29.57
3.	5.0	72.0	66.0	0.0	68.00	29.57
4.	5.8	72.0	66.0	0.0	68.00	29.56
5.	5.4	73.3	66.7	247.5	69.00	29.56
6.	5.0	74.7	67.3	947.6	70.00	29.56
7.	4.6	76.0	68.0	1788.6	71.00	29.56
8.	4.6	78.3	69.3	2296.4	72.33	29.55
9.	4.6	80.7	70.7	2556.1	73.67	29.55
10.	4.6	83.0	72.0	2522.9	75.00	29.54
11.	7.7	84.7	71.3	2725.0	75.33	29.53
12.	10.7	86.3	70.7	2794.0	75.67	29.52
13.	13.8	88.0	70.0	2725.0	76.00	29.51
14.	11.9	85.0	69.3	2287.7	74.67	29.51
15.	10.0	82.0	68.7	1772.2	73.33	29.52
16.	8.1	79.0	68.0	1238.5	72.00	29.52
17.	10.0	79.3	67.7	1042.2	71.67	29.53
18.	11.9	79.7	67.3	699.2	71.33	29.54
19.	13.8	80.0	67.0	247.6	71.00	29.55
20.	13.8	78.3	66.0	0.0	70.00	29.58
21.	13.8	76.7	65.0	0.0	69.00	29.60
22.	13.8	75.0	64.0	0.0	68.00	29.63
23.	10.4	73.0	62.0	0.0	66.00	29.65

********** METEOROLOGY FOR   7/ 4/72***

HOUR	WIND SP. (MPH)	DRY BULB (DEG.F)	DEWPOINT (DEG.F)	SOLAR RAD BTU/FT2/D	WET BULB (DEG.F)	ATM.PRESS. PSIA
0.	6.9	71.0	60.0	0.0	64.00	29.66
1.	3.5	69.0	58.0	0.0	62.00	29.68
2.	6.1	68.7	57.0	0.0	61.33	29.70
3.	8.8	68.3	56.0	0.0	60.67	29.71
4.	11.5	68.0	55.0	0.0	60.00	29.73
5.	10.0	67.7	55.0	264.8	60.00	29.75
6.	8.4	67.3	55.0	800.7	60.00	29.77
7.	6.9	67.0	55.0	1290.6	60.00	29.79
8.	8.4	69.3	55.3	1998.7	61.00	29.80
9.	10.0	71.7	55.7	2699.8	62.00	29.80
10.	11.5	74.0	56.0	3310.8	63.00	29.81
11.	10.0	74.7	55.7	3253.5	63.00	29.81
12.	8.4	75.3	55.3	2910.2	63.00	29.81
13.	6.9	76.0	55.0	2331.0	63.00	29.81
14.	7.3	77.0	54.7	2627.5	63.33	29.80
15.	7.7	78.0	54.3	2627.6	63.67	29.80
16.	8.1	79.0	54.0	2337.9	64.00	29.79
17.	6.9	78.0	53.7	1586.1	63.33	29.79
18.	5.8	77.0	53.3	862.5	62.67	29.78
19.	4.6	76.0	53.0	244.1	62.00	29.78
20.	4.2	74.7	55.0	0.0	62.67	29.78
21.	3.8	73.3	57.0	0.0	63.33	29.79
22.	3.5	72.0	59.0	0.0	64.00	29.79
23.	4.6	71.0	57.3	0.0	62.67	29.79

Figure 8.8   (Continued)

NOTE: Output for dates 7/5/72 through 8/19/72 not shown because of its length.

********** METEOROLOGY FOR 8/20/72**

HOUR	WIND SP. (MPH)	DRY BULB (DEG.F)	DEWPOINT (DEG.F)	SOLAR RAD BTU/FT2/D	WET BULB (DEG.F)	ATM.PRESS. PSIA
0.	1.5	63.7	58.7	0.0	61.00	29.71
1.	0.0	62.0	58.0	0.0	60.00	29.72
2.	1.9	61.3	57.7	0.0	59.33	29.72
3.	3.8	60.7	57.3	0.0	58.67	29.73
4.	5.8	60.0	57.0	0.0	58.00	29.73
5.	5.4	61.0	57.7	0.0	59.00	29.74
6.	5.0	62.0	58.3	537.2	60.00	29.76
7.	4.6	63.0	59.0	1296.6	61.00	29.77
8.	4.6	67.3	58.3	2004.3	62.00	29.77
9.	4.6	71.7	57.7	2612.0	63.00	29.78
10.	4.6	76.0	57.0	3078.4	64.00	29.78
11.	6.9	78.0	57.0	3371.5	65.00	29.77
12.	9.2	80.0	57.0	3471.5	66.00	29.75
13.	11.5	82.0	57.0	3371.5	67.00	29.74
14.	10.0	82.7	55.7	3078.4	66.33	29.74
15.	8.4	83.3	54.3	2612.0	65.67	29.73
16.	6.9	84.0	53.0	2004.3	65.00	29.73
17.	6.1	82.3	53.0	1296.6	64.67	29.73
18.	5.4	80.7	53.0	537.2	64.33	29.74
19.	4.6	79.0	53.0	0.0	64.00	29.74
20.	3.1	74.3	54.0	0.0	62.67	29.74
21.	1.5	69.7	55.0	0.0	61.33	29.75
22.	0.0	65.0	56.0	0.0	60.00	29.75
23.	0.0	64.0	55.7	0.0	59.33	29.76

********** METEOROLOGY FOR 8/21/72**

HOUR	WIND SP. (MPH)	DRY BULB (DEG.F)	DEWPOINT (DEG.F)	SOLAR RAD BTU/FT2/D	WET BULB (DEG.F)	ATM.PRESS. PSIA
0.	0.0	63.0	55.3	0.0	58.67	29.76
1.	0.0	62.0	55.0	0.0	58.00	29.77
2.	0.0	61.0	54.7	0.0	57.33	29.77
3.	0.0	60.0	54.3	0.0	56.67	29.77
4.	0.0	59.0	54.0	0.0	56.00	29.77
5.	0.0	59.7	55.0	0.0	57.00	29.78
6.	0.0	60.3	56.0	521.7	58.00	29.79
7.	0.0	61.0	57.0	1280.5	59.00	29.80
8.	1.9	65.7	58.7	1987.5	61.67	29.79
9.	3.8	70.3	60.3	2594.6	64.33	29.79
10.	5.8	75.0	62.0	3060.5	67.00	29.78
11.	5.0	77.7	60.7	3353.4	67.00	29.77
12.	4.2	80.3	59.3	3453.3	67.00	29.76
13.	3.5	83.0	58.0	3353.4	67.33	29.75
14.	4.6	83.3	58.3	3025.2	67.33	29.73
15.	5.8	83.7	58.7	2474.7	67.67	29.72
16.	6.9	84.0	59.0	1780.8	68.00	29.70
17.	6.5	81.7	60.0	1188.0	67.67	29.69
18.	6.1	79.3	61.0	497.6	67.33	29.68
19.	5.8	77.0	62.0	0.0	67.00	29.67
20.	6.5	76.0	61.7	0.0	66.33	29.67
21.	7.3	75.0	61.3	0.0	66.00	29.67
22.	8.1	74.0	61.0	0.0	65.00	29.67
23.	6.9	72.3	60.7	0.0	65.00	29.67

Figure 8.8   (Continued)

8-19

```
**********NUMBER OF CARDS PUNCHED = 588 *******************************************

********** THE MONTHLY AVERAGE VALUES FROM 5/ 1/73 TO END OF DATA **************
```

1973	*RMS WIND SPEED	*DRY BULB (DEG.F)	*DEWPOINT (DEG.F)	* SOLAR RADIATION*	*WET BULB (DEG.F)	*ATM.PRESS PSIG
MAY	8.91	57.38	47.16	1378.4	52.03	29.54
JUNE	6.12	72.79	62.67	1662.2	66.35	29.64
JULY	6.43	76.14	63.78	1884.9	68.19	29.63
AUGUST	5.90	75.38	64.59	1539.1	68.37	29.67
SEPTEMBER	7.37	67.87	55.93	1291.5	60.89	29.71
1974						
MAY	8.61	63.47	46.71	1648.8	54.71	29.58
JUNE	7.59	70.60	57.27	1686.6	62.61	29.60
JULY	7.54	77.27	59.89	1763.9	66.46	29.64
AUGUST	5.74	76.47	63.89	1377.3	68.34	29.71
SEPTEMBER	7.46	64.24	55.62	1182.6	59.29	29.70
1975						
MAY	6.65	64.74	55.96	1559.6	59.57	29.61
JUNE	7.61	70.57	62.24	1636.9	65.36	29.67
JULY	6.84	75.01	66.34	1746.9	69.36	29.65
AUGUST	6.75	75.12	66.77	1507.4	69.63	29.70
SEPTEMBER	7.31	62.82	56.25	1157.0	58.99	29.76

Figure 8.8   (Continued)

```
 1  5.0 74.0 65.7   0.0  68.67 29.59  3.5 72.0 66.0    0.0  68.00 29.58
 2  4.2 72.0 66.0   0.0  68.00 29.57  5.0 72.0 66.0    0.0  68.00 29.57
 3  5.8 72.0 66.0   0.0  68.00 29.56  5.4 73.3 66.7 247.5  69.00 29.56
 4  5.0 74.7 67.3 947.6  70.00 29.56  4.6 76.0 68.01788.6  71.00 29.56
 5  4.6 78.3 69.32296.8  72.33 29.55  4.6 80.7 70.72556.1  73.67 29.55
 6  4.6 83.0 72.02522.9  75.00 29.54  7.7 84.7 71.32725.0  75.33 29.53
 7 10.7 86.3 70.72794.0  75.67 29.52 13.8 88.0 70.02725.0  76.00 29.51
 8 11.9 85.0 69.32287.7  74.67 29.51 10.0 82.0 68.71772.2  73.33 29.52
 9  8.1 79.0 68.01238.5  72.00 29.52 10.0 79.3 67.71042.2  71.67 29.53
10 11.9 79.7 67.3 699.2  71.33 29.54 13.8 80.0 67.0 247.6  71.00 29.55
11 13.8 78.3 66.0   0.0  70.00 29.58 13.8 76.7 65.0    0.0  69.00 29.60
12 13.8 75.0 64.0   0.0  68.00 29.63 10.4 73.0 62.0    0.0  66.00 29.65
13  6.9 71.0 60.0   0.0  64.00 29.66  3.5 69.0 58.0    0.0  62.00 29.68
14  6.1 68.7 57.0   0.0  61.33 29.70  8.8 68.3 56.0    0.0  60.67 29.71
15 11.5 68.0 55.0   0.0  60.00 29.73 10.0 67.7 55.0 264.8  60.00 29.75
16  8.4 67.3 55.0 800.7  60.00 29.77  6.9 67.0 55.01290.6  60.00 29.79
17  8.4 69.3 55.31998.7  61.00 29.80 10.0 71.7 55.72699.8  62.00 29.80
18 11.5 74.0 56.03310.8  63.00 29.81 10.0 74.7 55.73253.5  63.00 29.81
19  8.4 75.3 55.32910.2  63.00 29.81  6.9 76.0 55.02331.0  63.00 29.81
20  7.3 77.0 54.72627.5  63.33 29.80  7.7 78.0 54.32627.6  63.67 29.80
21  8.1 79.0 54.02337.9  64.00 29.79  6.9 78.0 53.71586.1  63.33 29.79
22  5.8 77.0 53.3 862.5  62.67 29.78  4.6 76.0 53.0 244.1  62.00 29.78
23  4.2 74.7 55.0   0.0  62.67 29.78  3.8 73.3 57.0    0.0  63.33 29.79
24  3.5 72.0 59.0   0.0  64.00 29.79  4.6 71.0 57.3    0.0  62.67 29.79
25  5.8 70.0 55.7   0.0  61.33 29.79  6.9 69.0 54.0    0.0  60.00 29.79
26  5.8 66.3 54.0   0.0  59.00 29.79  4.6 63.7 54.0    0.0  58.00 29.79
27  3.5 61.0 54.0   0.0  57.00 29.79  5.4 60.3 53.0 123.6  56.33 29.80
28  7.3 59.7 52.0 392.2  55.67 29.81  9.2 59.0 51.0 660.7  55.00 29.82
29 10.4 59.3 52.0 910.9  55.67 29.82 11.5 59.7 53.01125.8  56.33 29.81
30 12.7 60.0 54.01290.7  57.00 29.81 11.9 60.0 54.01394.4  57.00 29.80
31 11.1 60.0 54.01429.7  57.00 29.80 10.4 60.0 54.01394.4  57.00 29.79
32  8.8 61.7 54.71290.7  58.00 29.78  7.3 63.3 55.31125.8  59.00 29.77
33  5.8 65.0 56.0 910.9  60.00 29.76  7.7 65.7 56.3 660.7  60.33 29.76
34  9.6 66.3 56.7 392.2  60.67 29.76 11.5 67.0 57.0 123.6  61.00 29.76
35 11.1 66.3 57.0   0.0  60.67 29.76 10.7 65.7 57.0    0.0  60.33 29.77
36 10.4 65.0 57.0   0.0  60.00 29.77  9.2 64.7 56.3    0.0  59.67 29.77
37  8.1 64.3 55.7   0.0  59.33 29.77  6.9 64.0 55.0    0.0  59.00 29.77
38  6.5 63.7 55.0   0.0  58.67 29.77  6.1 63.3 55.0    0.0  58.33 29.77
39  5.8 63.0 55.0   0.0  58.00 29.77  5.8 63.0 54.7 121.8  58.00 29.78
40  5.8 63.0 54.3 390.2  58.00 29.80  5.8 63.0 54.0 658.9  58.00 29.81
41  5.8 63.3 54.3 908.6  58.33 29.82  5.8 63.7 54.71123.3  58.67 29.83
42  5.8 64.0 55.01288.1  59.00 29.84  5.8 65.3 54.71391.7  59.33 29.84
43  5.8 66.7 54.31427.0  59.67 29.84  5.8 68.0 54.01391.7  60.00 29.84
44  5.4 69.0 54.01596.4  60.33 29.83  5.0 70.0 54.01642.6  60.67 29.83
45  4.6 71.0 54.01516.0  61.00 29.82  4.6 70.3 55.71162.7  61.67 29.82
46  4.6 69.7 57.3 725.1  62.33 29.82  4.6 69.0 59.0 237.2  63.00 29.82
47  4.2 67.3 58.0   0.0  61.67 29.83  3.8 65.7 57.0    0.0  60.33 29.84
48  3.5 64.0 56.0   0.0  59.00 29.85  3.5 63.3 56.3    0.0  59.00 29.85
49  3.5 62.7 56.7   0.0  59.00 29.85  3.5 62.0 57.0    0.0  59.00 29.85
50  3.5 61.0 56.0   0.0  58.00 29.85  3.5 60.0 55.0    0.0  57.00 29.86
51  3.5 59.0 54.0   0.0  56.00 29.86  4.6 59.7 54.7 294.5  56.67 29.87
52  5.8 60.3 55.3 721.4  57.33 29.89  6.9 61.0 56.0 656.3  58.00 29.90
53  6.9 64.0 57.31512.0  60.00 29.90  6.9 67.0 58.72452.9  62.00 29.91
54  6.9 70.0 60.03290.5  64.00 29.91  6.1 72.0 59.03234.5  64.00 29.90
55  5.4 74.0 58.02893.5  64.00 29.88  4.6 76.0 57.02317.4  64.00 29.87
56  4.6 76.3 56.32269.4  64.00 29.86  4.6 76.7 55.72082.8  64.00 29.85
57  4.6 77.0 55.01764.5  64.00 29.84  5.0 76.0 56.01277.9  64.00 29.84
```

Figure 8.8   (Continued)

58	5.4	75.0	57.0	755.9	64.00	29.85	5.8	74.0	58.0	233.8	64.00	29.85
59	3.8	71.7	57.7	0.0	63.00	29.85	1.9	69.3	57.3	0.0	62.00	29.85
60	0.0	67.0	57.0	0.0	61.00	29.85	1.2	66.3	57.0	0.0	60.67	29.85
61	2.3	65.7	57.0	0.0	60.33	29.86	3.5	65.0	57.0	0.0	60.00	29.86
62	3.5	63.7	56.3	0.0	59.33	29.86	3.5	62.3	55.7	0.0	58.67	29.87
63	3.5	61.0	55.0	0.0	58.00	29.87	3.8	62.0	56.3	337.0	59.00	29.88
64	4.2	63.0	57.7	1090.6	60.00	29.88	4.6	64.0	59.0	1820.2	61.00	29.89
65	4.6	65.3	59.3	2356.4	61.67	29.89	4.6	66.7	59.7	2603.8	62.33	29.89
66	4.6	68.0	60.0	2497.3	63.00	29.89	4.6	70.7	59.0	2575.7	63.33	29.88
67	4.6	73.3	58.0	2509.3	63.67	29.86	4.6	76.0	57.0	2312.6	64.00	29.85
68	7.3	74.0	58.0	2140.0	64.00	29.85	10.0	72.0	59.0	1865.6	64.00	29.86
69	12.7	70.0	60.0	1507.9	64.00	29.86	10.7	70.0	59.7	1215.5	63.67	29.86
70	8.8	70.0	59.3	784.6	63.33	29.86	6.9	70.0	59.0	258.9	63.00	29.86
71	5.8	68.3	59.0	0.0	62.33	29.87	4.6	66.7	59.0	0.0	61.67	29.88
72	3.5	65.0	59.0	0.0	61.00	29.89	2.3	64.0	58.0	0.0	60.00	29.89
73	1.2	63.0	57.0	0.0	59.00	29.90	0.0	62.0	56.0	0.0	58.00	29.90
74	1.5	61.7	56.3	0.0	58.33	29.89	3.1	61.3	56.7	0.0	58.67	29.89
75	4.6	61.0	57.0	0.0	59.00	29.88	4.6	61.3	57.3	116.5	59.33	29.89
76	4.6	61.7	57.7	384.1	59.67	29.90	4.6	62.0	58.0	651.8	60.00	29.91
77	5.0	64.0	59.3	901.2	61.33	29.91	5.4	66.0	60.7	1115.3	62.67	29.92
78	5.8	68.0	62.0	1279.7	64.00	29.92	8.1	70.7	63.7	1714.0	66.00	29.91
79	10.4	73.3	65.3	2073.8	68.00	29.89	12.7	76.0	67.0	2307.6	70.00	29.88
80	12.3	77.7	67.0	1871.2	70.67	29.87	11.9	79.3	67.0	1382.3	71.33	29.85
81	11.5	81.0	67.0	901.2	72.00	29.84	10.7	80.0	65.3	881.7	70.67	29.83
82	10.0	79.0	63.7	641.0	69.33	29.83	9.2	78.0	62.0	226.8	68.00	29.82
83	10.0	76.7	63.3	0.0	68.33	29.83	10.7	75.3	64.7	0.0	68.67	29.83
84	11.5	74.0	66.0	0.0	69.00	29.84	10.0	73.7	65.7	0.0	68.67	29.84
85	8.4	73.3	65.3	0.0	68.33	29.84	6.9	73.0	65.0	0.0	68.00	29.84
86	4.6	72.3	64.7	0.0	67.67	29.84	2.3	71.7	64.3	0.0	67.33	29.83
87	0.0	71.0	64.0	0.0	67.00	29.83	3.1	71.3	63.7	304.0	66.67	29.83
88	6.1	71.7	63.3	1071.9	66.33	29.83	9.2	72.0	63.0	1855.5	66.00	29.83
89	10.7	74.3	63.7	2567.4	67.33	29.83	12.3	76.7	64.3	3178.6	68.67	29.84
90	13.8	79.0	65.0	3647.7	70.00	29.84	13.8	81.3	65.7	3760.4	71.00	29.84
91	13.8	83.7	66.3	3295.5	72.00	29.83	13.8	86.0	67.0	2302.5	73.00	29.83
92	13.0	85.0	66.7	1866.8	72.67	29.82	12.3	84.0	66.3	1378.8	72.33	29.81
93	11.5	83.0	66.0	898.6	72.00	29.80	11.9	81.7	66.0	649.4	71.33	29.81
94	12.3	80.3	66.0	382.1	70.67	29.81	12.7	79.0	66.0	114.7	70.00	29.82
95	11.5	78.7	65.7	0.0	70.00	29.83	10.4	78.3	65.3	0.0	70.00	29.85
96	9.2	78.0	65.0	0.0	70.00	29.86	10.0	77.3	64.7	0.0	69.33	29.87
97	10.7	76.7	64.3	0.0	68.67	29.87	11.5	76.0	64.0	0.0	68.00	29.88
98	7.7	74.7	63.7	0.0	67.33	29.88	3.8	73.3	63.3	0.0	66.67	29.88
99	0.0	72.0	63.0	0.0	66.00	29.88	0.0	73.0	64.0	322.5	67.00	29.89
100	0.0	74.0	65.0	1085.6	68.00	29.91	0.0	75.0	66.0	1848.7	69.00	29.92
101	1.9	77.3	67.3	2559.8	70.67	29.92	3.8	79.7	68.7	3170.4	72.33	29.92
102	5.8	82.0	70.0	3639.0	74.00	29.92	6.5	84.0	69.3	3888.1	74.00	29.92
103	7.3	86.0	68.7	3847.5	74.00	29.91	8.1	88.0	68.0	3524.4	74.00	29.91
104	8.4	88.7	68.7	3123.9	74.67	29.90	8.8	89.3	69.3	2584.2	75.33	29.89
105	9.2	90.0	70.0	1960.8	76.00	29.88	8.1	89.0	70.3	1506.9	76.00	29.88
106	6.9	88.0	70.7	931.9	76.00	29.87	5.8	87.0	71.0	289.0	76.00	29.87
107	6.1	85.3	67.3	0.0	73.33	29.88	6.5	83.7	63.7	0.0	70.67	29.89
108	6.9	82.0	60.0	0.0	68.00	29.90	6.1	80.0	62.7	0.0	69.00	29.90
109	5.4	78.0	65.3	0.0	70.00	29.91	4.6	76.0	68.0	0.0	71.00	29.91
110	5.0	75.3	67.3	0.0	70.33	29.90	5.4	74.7	66.7	0.0	69.67	29.90
111	5.8	74.0	66.0	0.0	69.00	29.89	5.0	74.3	66.7	315.3	69.33	29.89
112	4.2	74.7	67.3	1051.5	69.67	29.90	3.5	75.0	68.0	1734.0	70.00	29.90
113	4.2	76.7	69.0	2240.6	71.33	29.90	5.0	78.3	70.0	2502.0	72.67	29.89
114	5.8	80.0	71.0	2473.8	74.00	29.89	5.4	81.3	70.7	2552.5	74.00	29.88
115	5.0	82.7	70.3	2486.9	74.00	29.86	4.6	84.0	70.0	2291.7	74.00	29.85
116	5.4	84.3	69.3	1857.8	73.67	29.83	6.1	84.7	68.7	1371.6	73.33	29.81
117	6.9	85.0	68.0	893.2	73.00	29.79	6.9	84.0	68.3	798.9	73.00	29.78
118	6.9	83.0	68.7	552.5	73.00	29.78	6.9	82.0	69.0	185.3	73.00	29.77

Figure 8.8 (Continued)

NOTE: Cards 119 to 427 not shown because of length of output.

```
428 11.1 79.3 58.32422.5   66.33  29.49 11.9 80.7 57.72221.6   66.67  29.49
429 12.7 82.0 57.01839.1   67.00  29.49  8.4 79.3 59.01329.4   67.00  29.48
430  4.2 76.7 61.0 677.3   67.00  29.48  0.0 74.0 63.0    0.0  67.00  29.47
431  2.3 74.0 62.7    0.0  66.67  29.48  4.6 74.0 62.3    0.0  66.33  29.49
432  6.9 74.0 62.0    0.0  66.00  29.50  4.6 72.7 62.3    0.0  66.00  29.50
433  2.3 71.3 62.7    0.0  66.00  29.49  0.0 70.0 63.0    0.0  66.00  29.49
434  0.0 68.7 62.7    0.0  65.00  29.48  0.0 67.3 62.3    0.0  64.00  29.47
435  0.0 66.0 62.0    0.0  63.00  29.46  0.0 67.0 63.0    0.0  64.00  29.47
436  0.0 68.0 64.0 500.9   65.00  29.48  0.0 69.0 65.0 695.4   66.00  29.49
437  2.7 72.7 65.71196.3   67.67  29.49  5.4 76.3 66.31725.3   69.33  29.49
438  8.1 80.0 67.02222.5   71.00  29.49 10.7 81.7 66.02197.5   71.00  29.49
439 13.4 83.3 65.02006.1   71.00  29.48 16.1 85.0 64.01683.8   71.00  29.48
440 15.3 84.0 62.31904.5   70.00  29.49 14.6 83.0 60.71902.8   69.00  29.51
441 13.8 82.0 59.01670.3   68.00  29.52 13.0 78.7 56.71162.0   65.33  29.55
442 12.3 75.3 54.3 574.2   62.67  29.58 11.5 72.0 52.0    0.0  60.00  29.61
443 10.7 69.3 52.3    0.0  59.33  29.64 10.0 66.7 52.7    0.0  58.67  29.67
444  9.2 64.0 53.0    0.0  58.00  29.70  6.1 63.0 52.7    0.0  57.33  29.71
445  3.1 62.0 52.3    0.0  56.67  29.73  0.0 61.0 52.0    0.0  56.00  29.74
446  2.7 60.7 51.7    0.0  55.67  29.75  5.4 60.3 51.3    0.0  55.33  29.76
447  8.1 60.0 51.0    0.0  55.00  29.77  8.4 60.3 51.0    0.0  55.00  29.79
448  8.8 60.7 51.0 687.4   55.00  29.80  9.2 61.0 51.01443.9   55.00  29.82
449 10.0 63.0 50.02151.1   55.67  29.83 10.7 65.0 49.02758.4   56.33  29.84
450 11.5 67.0 48.03224.3   57.00  29.85 11.5 68.7 47.33448.3   57.33  29.85
451 11.5 70.3 46.73427.9   57.67  29.84 11.5 72.0 46.03172.1   58.00  29.84
452 11.5 72.7 46.02849.3   58.33  29.84 11.5 73.3 46.02383.4   58.67  29.84
453 11.5 74.0 46.01813.3   59.00  29.84  8.8 73.0 47.01348.3   59.00  29.84
454  6.1 72.0 48.0 877.0   59.00  29.84  3.5 71.0 49.0    0.0  59.00  29.84
455  2.3 67.0 49.3    0.0  57.33  29.85  1.2 63.0 49.7    0.0  55.67  29.87
456  0.0 59.0 50.0    0.0  54.00  29.88  0.0 58.0 49.7    0.0  53.33  29.89
457  0.0 57.0 49.3    0.0  52.67  29.90  0.0 56.0 49.0    0.0  52.00  29.91
458  0.0 54.7 48.3    0.0  51.00  29.91  0.0 53.3 47.7    0.0  50.00  29.91
459  0.0 52.0 47.0    0.0  49.00  29.91  1.5 53.0 48.3    0.0  50.33  29.92
460  3.1 54.0 49.7 604.2   51.67  29.94  4.6 55.0 51.01101.5   53.00  29.95
461  4.2 59.0 52.01926.0   55.00  29.94  3.8 63.0 53.02688.8   57.00  29.94
462  3.5 67.0 54.03229.5   59.00  29.93  5.0 70.0 53.33483.5   60.00  29.91
463  6.5 73.0 52.73457.2   61.00  29.88  8.1 76.0 52.03157.7   62.00  29.86
464  8.1 76.7 53.02893.6   62.67  29.84  8.1 77.3 54.02473.5   63.33  29.83
465  8.1 78.0 55.01926.0   64.00  29.81  8.8 76.3 55.31353.8   63.67  29.80
466  9.6 74.7 55.7 656.8   63.33  29.79 10.4 73.0 56.0    0.0  63.00  29.78
467 10.0 72.0 56.3    0.0  62.67  29.78  9.6 71.0 56.7    0.0  62.33  29.78
468  9.2 70.0 57.0    0.0  62.00  29.78  6.1 69.0 57.0    0.0  61.67  29.78
469  3.1 68.0 57.0    0.0  61.33  29.78  0.0 67.0 57.0    0.0  61.00  29.78
470  0.0 66.7 57.3    0.0  61.00  29.77  0.0 66.3 57.7    0.0  61.00  29.75
471  0.0 66.0 58.0    0.0  61.00  29.74  0.0 66.0 58.3    0.0  61.33  29.75
472  0.0 66.0 58.7 407.3   61.67  29.75  0.0 66.0 59.0 497.9   62.00  29.76
473  0.0 66.0 59.7 746.9   62.33  29.76  0.0 66.0 60.3 960.6   62.67  29.77
474  0.0 66.0 61.01124.7   63.00  29.77  1.5 68.7 62.31521.7   64.67  29.76
475  3.1 71.3 63.71846.8   66.33  29.75  4.6 74.0 65.02048.7   68.00  29.74
476  5.0 75.7 66.31762.9   69.33  29.72  5.4 77.3 67.71404.7   70.67  29.70
477  5.8 79.0 69.01010.4   72.00  29.68  3.8 79.0 68.3 673.6   71.67  29.67
478  1.9 79.0 67.7 312.1   71.33  29.67  0.0 79.0 67.0    0.0  71.00  29.66
479  1.9 77.7 67.3    0.0  70.67  29.68  3.8 76.3 67.7    0.0  70.33  29.70
480  5.8 75.0 68.0    0.0  70.00  29.72  3.8 74.0 67.7    0.0  69.67  29.71
481  1.9 73.0 67.3    0.0  69.33  29.71  0.0 72.0 67.0    0.0  69.00  29.70
482  0.0 71.7 67.0    0.0  68.67  29.70  0.0 71.3 67.0    0.0  68.33  29.70
483  0.0 71.0 67.0    0.0  68.00  29.70  0.0 71.0 67.0    0.0  68.00  29.71
484  0.0 71.0 67.0 225.4   68.00  29.71  0.0 71.0 67.0 492.5   68.00  29.72
485  1.2 72.3 67.01236.9   68.67  29.72  2.3 73.7 67.02090.1   69.33  29.73
486  3.5 75.0 67.02864.6   70.00  29.72  4.6 77.3 67.32924.2   71.00  29.73
487  5.8 79.7 67.72751.5   72.00  29.72  6.9 82.0 68.02379.5   73.00  29.72
488  7.7 83.0 66.72529.7   72.33  29.71  8.4 84.0 65.32395.6   71.67  29.70
489  9.2 85.0 64.01994.1   71.00  29.69  7.7 83.7 62.71370.5   70.00  29.69
```

Figure 8.B   (Continued)

490	6.1	82.3	61.3	639.9	69.00	29.69	4.6	81.0	60.0	0.0	68.00	29.69
491	4.2	77.0	61.0	0.0	67.00	29.70	3.8	73.0	62.0	0.0	66.00	29.72
492	3.5	69.0	63.0	0.0	65.00	29.73	2.3	68.3	62.7	0.0	64.67	29.73
493	1.2	67.7	62.3	0.0	64.33	29.73	0.0	67.0	62.0	0.0	64.00	29.73
494	1.5	67.0	62.7	0.0	64.33	29.72	3.1	67.0	63.3	0.0	64.67	29.72
495	4.6	67.0	64.0	0.0	65.00	29.71	5.4	67.0	64.3	0.0	65.33	29.72
496	6.1	67.0	64.7	628.9	65.67	29.72	6.9	67.0	65.0	1391.5	66.00	29.73
497	6.9	71.3	65.7	2088.5	67.67	29.73	6.9	75.7	66.3	2641.8	69.33	29.73
498	6.9	80.0	67.0	2994.5	71.00	29.73	6.9	82.7	66.3	2983.1	71.67	29.71
499	6.9	85.3	65.7	2643.1	72.33	29.70	6.9	88.0	65.0	2029.4	73.00	29.68
500	8.4	88.7	64.7	2516.7	73.00	29.67	10.0	89.3	64.3	2516.5	73.00	29.65
501	11.5	90.0	64.0	2088.5	73.00	29.64	9.2	87.3	64.0	1291.0	72.00	29.63
502	6.9	84.7	64.0	497.6	71.00	29.63	4.6	82.0	64.0	0.0	70.00	29.62
503	6.1	80.0	64.3	0.0	69.67	29.62	7.7	78.0	64.7	0.0	69.33	29.61
504	9.2	76.0	65.0	0.0	69.00	29.61	8.4	75.3	64.3	0.0	68.33	29.61
505	7.7	74.7	63.7	0.0	67.67	29.62	6.9	74.0	63.0	0.0	67.00	29.62
506	7.7	74.3	63.7	0.0	67.67	29.62	8.4	74.7	64.3	0.0	68.33	29.63
507	9.2	75.0	65.0	0.0	69.00	29.63	10.0	72.7	62.0	0.0	66.33	29.67
508	10.7	70.3	59.0	266.2	63.67	29.71	11.5	68.0	56.0	651.5	61.00	29.75
509	12.3	70.0	55.0	1483.4	61.33	29.77	13.0	72.0	54.0	2367.0	61.67	29.79
510	13.8	74.0	53.0	3081.7	62.00	29.81	13.8	75.7	52.7	3395.7	62.33	29.81
511	13.8	77.3	52.3	3517.4	62.67	29.82	13.8	79.0	52.0	3435.7	63.00	29.82
512	13.0	78.3	52.0	3017.8	62.67	29.82	12.3	77.7	52.0	2366.9	62.33	29.83
513	11.5	77.0	52.0	1598.1	62.00	29.83	10.0	75.0	51.3	1053.9	61.00	29.84
514	8.4	73.0	50.7	470.1	60.00	29.86	6.9	71.0	50.0	0.0	59.00	29.87
515	4.6	68.0	51.0	0.0	58.33	29.88	2.3	65.0	52.0	0.0	57.67	29.89
516	0.0	62.0	53.0	0.0	57.00	29.90	0.0	61.3	53.0	0.0	56.67	29.91
517	0.0	60.7	53.0	0.0	56.33	29.91	0.0	60.0	53.0	0.0	56.00	29.92
518	0.0	59.7	53.0	0.0	56.00	29.92	0.0	59.3	53.0	0.0	56.00	29.93
519	0.0	59.0	53.0	0.0	56.00	29.93	0.0	60.0	54.0	0.0	56.67	29.94
520	0.0	61.0	55.0	502.9	57.33	29.96	0.0	62.0	56.0	1280.6	58.00	29.97
521	2.7	64.7	56.0	1889.1	59.33	29.96	5.4	67.3	56.0	2352.4	60.67	29.96
522	8.1	70.0	56.0	2635.8	62.00	29.95	7.7	70.7	54.0	2636.0	61.33	29.93
523	7.3	71.3	52.0	2413.5	60.67	29.91	6.9	72.0	50.0	2009.7	60.00	29.89
524	8.1	73.3	50.3	1838.0	60.33	29.87	9.2	74.7	50.7	1564.8	60.67	29.84
525	10.4	76.0	51.0	1208.9	61.00	29.82	9.2	74.0	51.0	794.3	60.33	29.81
526	8.1	72.0	51.0	349.5	59.67	29.80	6.9	70.0	51.0	0.0	59.00	29.79
527	7.7	69.3	51.0	0.0	58.67	29.80	8.4	68.7	51.0	0.0	58.33	29.80
528	9.2	68.0	51.0	0.0	58.00	29.81	9.2	67.0	52.3	0.0	58.33	29.79
529	9.2	66.0	53.7	0.0	58.67	29.76	9.2	65.0	55.0	0.0	59.00	29.74
530	8.4	64.3	55.7	0.0	59.00	29.72	7.7	63.7	56.3	0.0	59.00	29.71
531	6.9	63.0	57.0	0.0	59.00	29.69	4.6	63.3	57.7	0.0	59.67	29.69
532	2.3	63.7	58.3	204.1	60.33	29.68	0.0	64.0	59.0	470.5	61.00	29.68
533	2.3	64.3	59.3	718.8	61.33	29.68	4.6	64.7	59.7	932.0	61.67	29.67
534	6.9	65.0	60.0	1095.6	62.00	29.67	4.6	65.3	61.0	1198.4	62.67	29.66
535	2.3	65.7	62.0	1233.5	63.33	29.65	0.0	66.0	63.0	1198.4	64.00	29.64
536	0.0	66.7	63.0	1095.6	64.33	29.63	0.0	67.3	63.0	932.0	64.67	29.61
537	0.0	68.0	63.0	718.8	65.00	29.60	1.9	67.3	63.3	470.5	65.00	29.60
538	3.8	66.7	63.7	204.1	65.00	29.60	5.8	66.0	64.0	0.0	65.00	29.60
539	6.1	65.7	63.7	0.0	64.67	29.60	6.5	65.3	63.3	0.0	64.33	29.60
540	6.9	65.0	63.0	0.0	64.00	29.60	4.6	65.0	62.7	0.0	63.67	29.60
541	2.3	65.0	62.3	0.0	63.33	29.60	0.0	65.0	62.0	0.0	63.00	29.60
542	0.0	65.0	62.0	0.0	63.00	29.59	0.0	65.0	62.0	0.0	63.00	29.59
543	0.0	65.0	62.0	0.0	63.00	29.58	0.0	65.3	62.3	0.0	63.33	29.59
544	0.0	65.7	62.7	198.7	63.67	29.59	0.0	66.0	63.0	465.0	64.00	29.60
545	0.0	67.7	63.7	713.1	65.00	29.60	0.0	69.3	64.3	926.1	66.00	29.61
546	0.0	71.0	65.0	1089.6	67.00	29.61	2.3	74.3	66.7	2105.1	69.00	29.59
547	4.6	77.7	68.3	2858.5	71.00	29.58	6.9	81.0	70.0	3207.4	73.00	29.56
548	8.1	83.3	70.7	3032.2	74.33	29.54	9.2	85.7	71.3	2628.8	75.67	29.52
549	10.4	88.0	72.0	2037.4	77.00	29.50	8.8	86.3	72.0	1313.2	76.33	29.50
550	7.3	84.7	72.0	541.6	75.67	29.51	5.8	83.0	72.0	0.0	75.00	29.51
551	5.4	81.3	71.0	0.0	74.00	29.52	5.0	79.7	70.0	0.0	73.00	29.52

Figure 8.8 (Continued)

```
552   4.6 78.0 69.0    0.0   72.00  29.53   4.6 76.3 68.3    0.0   71.00  29.54
553   4.6 74.7 67.7    0.0   70.00  29.55   4.6 73.0 67.0    0.0   69.00  29.56
554   3.1 70.3 65.3    0.0   67.00  29.57   1.5 67.7 63.7    0.0   65.00  29.57
555   0.0 65.0 62.0    0.0   63.00  29.58   2.7 67.7 62.7    0.0   64.33  29.60
556   5.4 70.3 63.3  546.1   65.67  29.62   8.1 73.0 64.01278.5  67.00  29.64
557   7.7 75.3 64.01927.5   68.00  29.65   7.3 77.7 64.02439.2  69.00  29.66
558   6.9 80.0 64.02773.9   70.00  29.67   7.7 80.7 62.32681.8  69.33  29.66
559   8.4 81.3 60.72269.5   68.67  29.66   9.2 82.0 59.01604.8  68.00  29.65
560   7.3 83.0 60.31807.9   69.00  29.64   5.4 84.0 61.71791.8  70.00  29.62
561   3.5 85.0 63.01548.0   71.00  29.61   3.8 83.3 61.31176.1  69.33  29.62
562   4.2 81.7 59.7  538.1   67.67  29.62   4.6 80.0 58.0    0.0   66.00  29.63
563   4.6 75.7 58.7    0.0   65.00  29.65   4.6 71.3 59.3    0.0   64.00  29.67
564   4.6 67.0 60.0    0.0   63.00  29.69   3.1 65.3 59.3    0.0   62.00  29.70
565   1.5 63.7 58.7    0.0   61.00  29.71   0.0 62.0 58.0    0.0   60.00  29.72
566   1.9 61.3 57.7    0.0   59.33  29.72   3.8 60.7 57.3    0.0   58.67  29.73
567   5.8 60.0 57.0    0.0   58.00  29.73   5.4 61.0 57.7    0.0   59.00  29.74
568   5.0 62.0 58.3  537.2   60.00  29.76   4.6 63.0 59.01296.6  61.00  29.77
569   4.6 67.3 58.32004.3   62.00  29.77   4.6 71.7 57.72612.0  63.00  29.78
570   4.6 76.0 57.03078.4   64.00  29.78   6.9 78.0 57.03371.5  65.00  29.77
571   9.2 80.0 57.03471.5   66.00  29.75  11.5 82.0 57.03371.5  67.00  29.74
572  10.0 82.7 55.73078.4   66.33  29.74   8.4 83.3 54.32612.0  65.67  29.73
573   6.9 84.0 53.02004.3   65.00  29.73   6.1 82.3 53.01296.6  64.67  29.73
574   5.4 80.7 53.0  537.2   64.33  29.74   4.6 79.0 53.0    0.0   64.00  29.74
575   3.1 74.3 54.0    0.0   62.67  29.74   1.5 69.7 55.0    0.0   61.33  29.75
576   0.0 65.0 56.0    0.0   60.00  29.75   0.0 64.0 55.7    0.0   59.33  29.76
577   0.0 63.0 55.3    0.0   58.67  29.76   0.0 62.0 55.0    0.0   58.00  29.77
578   0.0 61.0 54.7    0.0   57.33  29.77   0.0 60.0 54.3    0.0   56.67  29.77
579   0.0 59.0 54.0    0.0   56.00  29.77   0.0 59.7 55.0    0.0   57.00  29.78
580   0.0 60.3 56.0  521.7   58.00  29.79   0.0 61.0 57.01280.5  59.00  29.80
581   1.9 65.7 58.71987.5   61.67  29.79   3.8 70.3 60.32594.6  64.33  29.79
582   5.8 75.0 62.03060.5   67.00  29.78   5.0 77.7 60.73353.4  67.00  29.77
583   4.2 80.3 59.33453.3   67.00  29.76   3.5 83.0 58.03353.4  67.00  29.75
584   4.6 83.3 58.33025.2   67.33  29.73   5.8 83.7 58.72474.7  67.67  29.72
585   6.9 84.0 59.01780.8   68.00  29.70   6.5 81.7 60.01188.0  67.67  29.69
586   6.1 79.3 61.0  497.6   67.33  29.68   5.8 77.0 62.0    0.0   67.00  29.67
587   6.5 76.0 61.7    0.0   66.67  29.67   7.3 75.0 61.3    0.0   66.33  29.67
588   8.1 74.0 61.0    0.0   66.00  29.67   6.9 72.3 60.7    0.0   65.00  29.67
```

Figure 8.8   (Continued)

The input for run 2 is also shown in Figure 8.9. The parameter IMET = 1 specifies that the fixed values are used for meteorology for the run, namely TA = 90, TW = 70, TD = 60.1, W = 3, and HS = 1500. The parameter ISPRAY = 1 specifies the regression model. IEVAP = 0 specifies that the pond remains full during the run. The parameter TSPRON = 200 specifies that the sprays are off until 200 hr into the run. Essentially, this allows the pond temperature to reach equilibrium before the effects of the sprays are felt, allowing a more-accurate prediction of the peak temperature attributable to heat load alone.

```
-.60637276E+00    .40195127E-03    .38449863E-02    .18230236E-02
-.34078270E-01    .30138737E+00   -.25690451E+01    .65576685E-01
-.73791051E-03    .26319278E-05    .35669730E-02    .12911864E-01
-.39275022E-04   -.41450389E-01    .14646531E-03   -.33234415E-03
 .41560445E-03   -.12268707E-02    .11416664E-01   -.86122112E-01
 .28767122E-02   -.29725976E-04    .10168749E-06   -.27394599E-03
 .28406611E-04    .22034012E-05
$PARAM WID=183,ALEN=283,HT=12.0,THETA=71.0,VELO=22.47,R=.095,
   YO=5.0,WDRO=0,NDRIFT=6,DWDR=10,FDRIFT=.0005,.00058,.001914,.004346,
   .00789,.0143303$
1176
$HFT NH=14.,TH=0.,.01,1.,1.1,1.9,3.9,5.,8.,12.,24.,29.,140.,840.,2000.,
   HEAT(1)=0.,0.,.85E9,2*.51E9,.5E9,.68E9,.6E9,.4E9,.31E9,.27E9,
   .21E9,.18E9,.1E9,FLOW=14*205200.0$
   EFFECTS OF DESIGN BASIS METEOROLOGY WITH STEADY HEAT LOAD-SPRAYS ON
$INLIST VZERO=2942357,A=422000,NSTEPS=2000,NPRINT=10,TZERO=80,
   IMET=0,ISPRAY=1,TSKIP=5000,Q1=0.23E9,F1=2.052E5,TSPRON=0.0,IEVAP=0$
   EFFECTS OF DESIGN BASIS HEAT LOAD ONLY
$INLIST VZERO=2942357.,A=422000.,NSTEPS=2000,NPRINT=10,TZERO=90,IMET=1,
   TSKIP=200,DT=.5,TA=90,TW=70,TD=60.1,W=0,HS=1500,ISPRAY=1,IEVAP=0,
   TSPRON=200,Q1=0,F1=1$
```

Figure 8.9   Input deck for program SPRND, procedure 1

The output from run 2 is shown printed in Figure 8.12 and plotted in
Figure 8.13.

Run 3 would be set up after inspection of runs 1 and 2. The peak temperature
for ambient meteorology and steady heat load occurred at 451.0 hr. The peak
temperature for the design-basis heat load alone occurred at 213.0 hr, or
15.0 hr after the sprays were turned on. The parameter TSKIP should, therefore,
be 451.0 hr - 13.0 hr = 438.0 hr.

The data input for run 3 is shown in Figure 8.14. The parameter TSPRON = 438.0 hr
delays the sprays 438.0 hr. The parameters Q1 = 0.0 and F1 = 1.0 specify that
the heat load and flowrate to the pond are 0 Btu/hr and 1 ft^3/hr for times less
than 438.0 hr. The parameter IMET = 0 specifies that the meteorological table
is used for input. IEVAP = 1 specifies that the pond volume is allowed to
change in response to water loss for times greater than 438.0 hr. The parameter
ISPRAY = 2 specifies that the rigorous spray model is used.

EFFECTS OF DESIGN BASIS METEOROLOGY WITH STEADY HEAT LOAD-SPRAYS ON

SPRAY FIELD PARAMETERS
****
INITIAL VELOCITY OF DROPS LEAVING NOZZLE, VELO = 684.89 CM/SEC
INITIAL ANGLE OF DROPS TO HOR., THETA = 1.239 RADIANS
GEOMETRIC MEAN RADIUS OF DROPS, R = .0950 CM
HEIGHT OF SPRAY FIELD, HT = 365.76 CM
WIDTH OF SPRAY FIELD, WID = 5577.8 CM
LENGTH OF SPRAY FIELD, ALEN = 8625.8 CM
HEIGHT OF SPRAY NOZZLES ABOVE POND SURFACE, YO = 152.4
HEADING OF WIND W.R.T.LONG AXIS, PHI = 90.00 DEGREES

POND PARAMETERS
****
INITIAL POND VOLUME,VZERO = 2942357.0 CU.FT.
POND SURFACE AREA,A = 422000.0 SQ.FT.
BLOWDOWN AND LEAKAGE,BLOW = 0.00 CU.FT./HR.
NUMBER OF INTEGRATION STEPS,NSTEPS = 2000
PRINT INTERVAL,NPRINT = 10
INTEGRATION TIMESTEP,DT = .50 HOURS
INITIAL POND TEMPERATURE,TZERO = 80.00 DEG.F
DELAY FOR HEAT TABLE,TSKIP = 5000.00 HRS
BASE HEAT LOAD ADDED TO TABLE,QBASE = 0.00 HRS
BASE FLOW RATE ADDED TO TABLE ,FBASE = 0.   CU.FT./HR.

HEAT IN BTU/HR	TIME FROM START	FLOW IN FT**3/HR
0.	0.00	.205E+06
0.	.01	.205E+06
.850E+09	1.00	.205E+06
.510E+09	1.10	.205E+06
.510E+09	1.90	.205E+06
.500E+09	3.90	.205E+06
.680E+09	5.00	.205E+06
.600E+09	8.00	.205E+06
.400E+09	12.00	.205E+06
.310E+09	24.00	.205E+06
.270E+09	29.00	.205E+06
.210E+09	140.00	.205E+06
.180E+09	840.00	.205E+06
.100E+09	2000.00	.205E+06

FOR TIME LESS THAN TSKIP
Q1 = .230E+09 BTU/HR
F1 = .205E+06 FT**3/HR

Figure 8.10 Output from program SPRPND, effect of ambient meteorology and steady heat load

METEOROLOGICAL TABLE USED AS INPUT

REGRESSION EQUATIONS USED FOR SPRAY MODEL

SPRAYS WILL BE DELAYED   0.00 HOURS

....

*************** MODEL RESULTS ***************

..TIME.......	TEMPERATURE (F)	....	VOLUME....
: HR		:	FT**3 :
5.00	80.58		.29423570E+07
10.00	82.48		.29423570E+07
15.00	84.21		.29423570E+07
20.00	83.76		.29423570E+07
25.00	82.27		.29423570E+07
30.00	80.57		.29423570E+07
35.00	80.35		.29423570E+07
40.00	81.02		.29423570E+07
45.00	80.92		.29423570E+07
50.00	80.29		.29423570E+07
55.00	79.26		.29423570E+07
60.00	77.43		.29423570E+07
65.00	77.21		.29423570E+07
70.00	76.45		.29423570E+07
75.00	76.27		.29423570E+07
80.00	76.53		.29423570E+07
85.00	77.21		.29423570E+07
90.00	78.04		.29423570E+07
95.00	78.22		.29423570E+07
100.00	78.14		.29423570E+07
105.00	78.26		.29423570E+07
110.00	79.78		.29423570E+07
115.00	80.45		.29423570E+07
120.00	80.28		.29423570E+07
125.00	79.77		.29423570E+07
130.00	80.28		.29423570E+07
135.00	81.01		.29423570E+07
140.00	80.31		.29423570E+07
145.00	79.97		.29423570E+07
150.00	79.59		.29423570E+07
155.00	79.82		.29423570E+07
160.00	80.66		.29423570E+07
165.00	80.78		.29423570E+07
170.00	80.71		.29423570E+07
175.00	81.15		.29423570E+07
180.00	81.95		.29423570E+07
185.00	82.56		.29423570E+07

Figure 8.10   (Continued)

190.00	82.05	.29423570E+07
195.00	81.58	.29423570E+07
200.00	82.49	.29423570E+07
205.00	84.93	.29423570E+07
210.00	86.44	.29423570E+07
215.00	86.17	.29423570E+07
220.00	85.40	.29423570E+07
225.00	85.54	.29423570E+07
230.00	86.79	.29423570E+07
235.00	86.92	.29423570E+07
240.00	86.13	.29423570E+07
245.00	85.28	.29423570E+07
250.00	84.92	.29423570E+07
255.00	85.88	.29423570E+07
260.00	84.09	.29423570E+07
265.00	82.87	.29423570E+07
270.00	82.51	.29423570E+07
275.00	84.23	.29423570E+07
280.00	85.54	.29423570E+07
285.00	85.72	.29423570E+07
290.00	85.18	.29423570E+07
295.00	84.90	.29423570E+07
300.00	86.40	.29423570E+07
305.00	87.24	.29423570E+07
310.00	86.33	.29423570E+07
315.00	85.60	.29423570E+07
320.00	85.85	.29423570E+07
325.00	87.22	.29423570E+07
330.00	87.17	.29423570E+07
335.00	86.62	.29423570E+07
340.00	86.13	.29423570E+07
345.00	85.78	.29423570E+07
350.00	87.40	.29423570E+07
355.00	87.78	.29423570E+07
360.00	87.33	.29423570E+07
365.00	86.63	.29423570E+07
370.00	87.05	.29423570E+07
375.00	88.70	.29423570E+07
380.00	89.38	.29423570E+07
385.00	88.90	.29423570E+07
390.00	88.23	.29423570E+07
395.00	89.42	.29423570E+07
400.00	91.39	.29423570E+07
405.00	91.36	.29423570E+07
410.00	90.52	.29423570E+07
415.00	89.55	.29423570E+07
420.00	90.57	.29423570E+07
425.00	90.99	.29423570E+07
430.00	90.98	.29423570E+07
435.00	90.05	.29423570E+07
440.00	89.91	.29423570E+07
445.00	91.29	.29423570E+07
450.00	92.12	.29423570E+07
455.00	91.68	.29423570E+07
460.00	90.33	.29423570E+07

Figure 8.10   (Continued)

465.00	89.66	.29423570E+07
470.00	90.70	.29423570E+07
475.00	92.11	.29423570E+07
480.00	91.39	.29423570E+07
485.00	90.05	.29423570E+07
490.00	90.31	.29423570E+07
495.00	91.18	.29423570E+07
500.00	91.07	.29423570E+07
505.00	89.51	.29423570E+07
510.00	88.44	.29423570E+07
515.00	88.74	.29423570E+07
520.00	88.99	.29423570E+07
525.00	88.08	.29423570E+07
530.00	87.27	.29423570E+07
535.00	86.77	.29423570E+07
540.00	86.70	.29423570E+07
545.00	86.32	.29423570E+07
550.00	84.34	.29423570E+07
555.00	81.94	.29423570E+07
560.00	80.57	.29423570E+07
565.00	80.21	.29423570E+07
570.00	79.80	.29423570E+07
575.00	79.82	.29423570E+07
580.00	80.08	.29423570E+07
585.00	80.51	.29423570E+07
590.00	81.21	.29423570E+07
595.00	81.62	.29423570E+07
600.00	81.66	.29423570E+07
605.00	81.38	.29423570E+07
610.00	81.90	.29423570E+07
615.00	82.72	.29423570E+07
620.00	81.93	.29423570E+07
625.00	81.32	.29423570E+07
630.00	80.46	.29423570E+07
635.00	80.39	.29423570E+07
640.00	81.45	.29423570E+07
645.00	81.81	.29423570E+07
650.00	81.58	.29423570E+07
655.00	81.16	.29423570E+07
660.00	81.41	.29423570E+07
665.00	81.71	.29423570E+07
670.00	81.85	.29423570E+07
675.00	81.38	.29423570E+07
680.00	81.41	.29423570E+07
685.00	81.94	.29423570E+07
690.00	82.67	.29423570E+07
695.00	82.95	.29423570E+07
700.00	82.48	.29423570E+07
705.00	82.36	.29423570E+07
710.00	83.71	.29423570E+07
715.00	84.40	.29423570E+07
720.00	84.23	.29423570E+07
725.00	83.79	.29423570E+07
730.00	84.51	.29423570E+07
735.00	85.48	.29423570E+07

Figure 8.10   (Continued)

740.00	85.28	.29423570E+07
745.00	84.68	.29423570E+07
750.00	84.30	.29423570E+07
755.00	84.74	.29423570E+07
760.00	86.18	.29423570E+07
765.00	86.39	.29423570E+07
770.00	85.88	.29423570E+07
775.00	85.05	.29423570E+07
780.00	84.19	.29423570E+07
785.00	82.30	.29423570E+07
790.00	80.78	.29423570E+07
795.00	80.07	.29423570E+07
800.00	79.65	.29423570E+07
805.00	80.32	.29423570E+07
810.00	81.21	.29423570E+07
815.00	80.95	.29423570E+07
820.00	80.53	.29423570E+07
825.00	80.48	.29423570E+07
830.00	81.46	.29423570E+07
835.00	82.01	.29423570E+07
840.00	81.72	.29423570E+07
845.00	81.19	.29423570E+07
850.00	81.28	.29423570E+07
855.00	81.34	.29423570E+07
860.00	81.57	.29423570E+07
865.00	81.53	.29423570E+07
870.00	81.60	.29423570E+07
875.00	82.51	.29423570E+07
880.00	82.31	.29423570E+07
885.00	80.35	.29423570E+07
890.00	79.35	.29423570E+07
895.00	78.06	.29423570E+07
900.00	77.47	.29423570E+07
905.00	77.17	.29423570E+07
910.00	77.32	.29423570E+07
915.00	77.63	.29423570E+07
920.00	78.00	.29423570E+07
925.00	79.18	.29423570E+07
930.00	79.76	.29423570E+07
935.00	78.91	.29423570E+07
940.00	79.11	.29423570E+07
945.00	79.64	.29423570E+07
950.00	80.83	.29423570E+07
955.00	82.21	.29423570E+07
960.00	82.68	.29423570E+07
965.00	82.98	.29423570E+07
970.00	83.65	.29423570E+07
975.00	85.00	.29423570E+07
980.00	84.90	.29423570E+07
985.00	83.97	.29423570E+07
990.00	83.14	.29423570E+07
995.00	83.81	.29423570E+07
1000.00	85.17	.29423570E+07

TSKIP = 5000.0 HOURS   MAX MODELED TEMPERATURE = 92.17 AT 451.00 HOURS

Figure 8.10   (Continued)

8-31

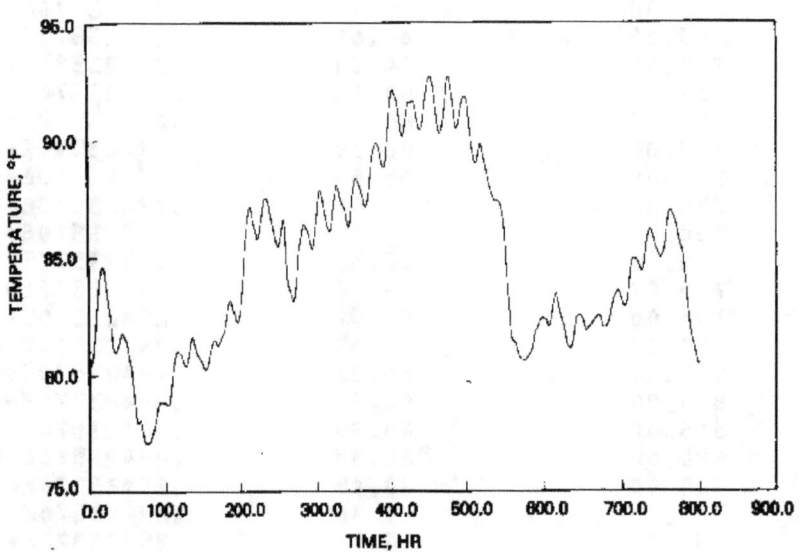

Figure 8.11   Pond temperature in response to a steady heat load and
ambient meteorology

The reliability of this estimate of the parameter ISKIP is questionable however, because of nonlinearities in the models that make the use of linear superposition strictly invalid. Therefore, a series of runs should be made varying TSTART over a range of 1 or 2 days. The results of this run will, therefore, not be shown, in favor of procedure 2 below.

8.6.2 Procedure 2

An alternative procedure for determining TSKIP is simply to vary this parameter over a wide range in a repetitive manner within the 50-day period of data and pick the value giving the highest pond temperature. This "brute force" approach is not particularly wasteful of computer time if the regression spray model is used for pond performance. The rigorous spray model may then be run for the value of TSKIP determined to give the highest pond temperature.

The input for the first run in procedure 2 is shown in Figure 8.15. The parameters IEVAP = 1 and IMET = 0 specify normal water loss and meteorological

EFFECTS OF DESIGN BASIS HEAT LOAD ONLY

SPRAY FIELD PARAMETERS
****
INITIAL VELOCITY OF DROPS LEAVING NOZZLE, VELO =   684.89 CM/SEC
INITIAL ANGLE OF DROPS TO HOR., THETA =   1.239 RADIANS
GEOMETRIC MEAN RADIUS OF DROPS, R =   .0950 CM
HEIGHT OF SPRAY FIELD, HT =   365.76 CM
WIDTH OF SPRAY FIELD, WID =   5577.8 CM
LENGTH OF SPRAY FIELD, ALEN =   8625.8 CM
HEIGHT OF SPRAY NOZZLES ABOVE POND SURFACE, YO =   152.4
HEADING OF WIND W.R.T.LONG AXIS, PHI =   90.00 DEGREES

POND PARAMETERS
****
INITIAL POND VOLUME,VZERO =   2942357.0 CU.FT.
POND SURFACE AREA,A =   422000.0 SQ.FT.
BLOWDOWN AND LEAKAGE,BLOW =   0.00 CU.FT./HR.
NUMBER OF INTEGRATION STEPS,NSTEPS = 1600
PRINT INTERVAL,NPRINT =   10
INTEGRATION TIMESTEP,DT =   .50 HOURS
INITIAL POND TEMPERATURE,TZERO =   90.00 DEG.F
DELAY FOR HEAT TABLE,TSKIP =   200.00 HRS
BASE HEAT LOAD ADDED TO TABLE,QBASE =   0.00 HRS
BASE FLOW RATE ADDED TO TABLE ,FBASE =   0.   CU.FT./HR.

| HEAT IN | TIME FROM | FLOW IN |
BTU/HR	START	FT**3/HR
0.	0.00	.205E+06
0.	.01	.205E+06
.850E+09	1.00	.205E+06
.510E+09	1.10	.205E+06
.510E+09	1.90	.205E+06
.500E+09	3.90	.205E+06
.680E+09	5.00	.205E+06
.600E+09	8.00	.205E+06
.400E+09	12.00	.205E+06
.310E+09	24.00	.205E+06
.270E+09	29.00	.205E+06
.210E+09	140.00	.205E+06
.180E+09	840.00	.205E+06
.100E+09	2000.00	.205E+06

Figure 8.12   Output from program SPRND, effect of design-basis heat load
and steady meteorology

```
FOR TIME LESS THAN TSKIP
Q1 =      0.       BTU/HR
F1 =      .100E+01 FT**3/HR
```

FIXED METEOROLOGICAL VALUES USED AS INPUT

```
DRY BULB TEMPERATURE,TA =        90.00 DEG. F
WET BULB TEMPERATURE,TW =        70.00 DEG. F
WIND SPEED,W =         0.00MPH
DEW POINT TEMPERATURE,TD =       60.10 DEG. F
SOLAR RADIATION,HS =     1500.00 BTU/SQ.FT./DAY
3AROMETRIC PRESSURE,PB =        29.92 IN.HG.
```

REGRESSION EGUATIONS USED FOR SPRAY MODEL

SPRAYS WILL BE DELAYED   200.00 HOURS

....

************* MODEL RESULTS *************

| ..TIME.. | TEMPERATURE (F) | ..VOLUME.. |
HR		FT**3
5.00	90.04	.29423570E+07
10.00	90.07	.29423570E+07
15.00	90.11	.29423570E+07
20.00	90.14	.29423570E+07
25.00	90.16	.29423570E+07
30.00	90.19	.29423570E+07
35.00	90.21	.29423570E+07
40.00	90.24	.29423570E+07
45.00	90.26	.29423570E+07
50.00	90.28	.29423570E+07
55.00	90.30	.29423570E+07
60.00	90.31	.29423570E+07
65.00	90.33	.29423570E+07
70.00	90.34	.29423570E+07
75.00	90.36	.29423570E+07
80.00	90.37	.29423570E+07
85.00	90.38	.29423570E+07
90.00	90.39	.29423570E+07
95.00	90.40	.29423570E+07
100.00	90.41	.29423570E+07
105.00	90.42	.29423570E+07
110.00	90.43	.29423570E+07
115.00	90.44	.29423570E+07
120.00	90.44	.29423570E+07
125.00	90.45	.29423570E+07

Figure 8.12   (Continued)

130.00	90.46	.29423570E+07
135.00	90.46	.29423570E+07
140.00	90.47	.29423570E+07
145.00	90.47	.29423570E+07
150.00	90.48	.29423570E+07
155.00	90.48	.29423570E+07
160.00	90.49	.29423570E+07
165.00	90.49	.29423570E+07
170.00	90.49	.29423570E+07
175.00	90.50	.29423570E+07
180.00	90.50	.29423570E+07
185.00	90.50	.29423570E+07
190.00	90.50	.29423570E+07
195.00	90.51	.29423570E+07
200.00	90.38	.29423570E+07
205.00	92.89	.29423570E+07
210.00	95.00	.29423570E+07
215.00	95.21	.29423570E+07
220.00	94.96	.29423570E+07
225.00	94.55	.29423570E+07
230.00	93.96	.29423570E+07
235.00	93.42	.29423570E+07
240.00	92.97	.29423570E+07
245.00	92.60	.29423570E+07
250.00	92.29	.29423570E+07
255.00	92.03	.29423570E+07
260.00	91.81	.29423570E+07
265.00	91.62	.29423570E+07
270.00	91.46	.29423570E+07
275.00	91.31	.29423570E+07
280.00	91.18	.29423570E+07
285.00	91.06	.29423570E+07
290.00	90.95	.29423570E+07
295.00	90.86	.29423570E+07
300.00	90.77	.29423570E+07
305.00	90.68	.29423570E+07
310.00	90.60	.29423570E+07
315.00	90.53	.29423570E+07
320.00	90.46	.29423570E+07
325.00	90.39	.29423570E+07
330.00	90.33	.29423570E+07
335.00	90.27	.29423570E+07
340.00	90.21	.29423570E+07
345.00	90.16	.29423570E+07
350.00	90.11	.29423570E+07
355.00	90.07	.29423570E+07
360.00	90.04	.29423570E+07
365.00	90.01	.29423570E+07
370.00	89.98	.29423570E+07
375.00	89.96	.29423570E+07
380.00	89.94	.29423570E+07
385.00	89.92	.29423570E+07
390.00	89.90	.29423570E+07
395.00	89.88	.29423570E+07

Figure 8.12   (Continued)

400.00	89.86	.29423570E+07
405.00	89.84	.29423570E+07
410.00	89.82	.29423570E+07
415.00	89.81	.29423570E+07
420.00	89.79	.29423570E+07
425.00	89.78	.29423570E+07
430.00	89.76	.29423570E+07
435.00	89.75	.29423570E+07
440.00	89.73	.29423570E+07
445.00	89.72	.29423570E+07
450.00	89.71	.29423570E+07
455.00	89.69	.29423570E+07
460.00	89.68	.29423570E+07
465.00	89.67	.29423570E+07
470.00	89.65	.29423570E+07
475.00	89.64	.29423570E+07
480.00	89.63	.29423570E+07
485.00	89.62	.29423570E+07
490.00	89.61	.29423570E+07
495.00	89.59	.29423570E+07
500.00	89.58	.29423570E+07
505.00	89.57	.29423570E+07
510.00	89.56	.29423570E+07
515.00	89.55	.29423570E+07
520.00	89.54	.29423570E+07
525.00	89.53	.29423570E+07
530.00	89.52	.29423570E+07
535.00	89.51	.29423570E+07
540.00	89.50	.29423570E+07
545.00	89.49	.29423570E+07
550.00	89.48	.29423570E+07
555.00	89.47	.29423570E+07
560.00	89.46	.29423570E+07
565.00	89.45	.29423570E+07
570.00	89.44	.29423570E+07
575.00	89.44	.29423570E+07
580.00	89.43	.29423570E+07
585.00	89.42	.29423570E+07
590.00	89.41	.29423570E+07
595.00	89.40	.29423570E+07
600.00	89.39	.29423570E+07
605.00	89.38	.29423570E+07
610.00	89.38	.29423570E+07
615.00	89.37	.29423570E+07
620.00	89.36	.29423570E+07
625.00	89.35	.29423570E+07
630.00	89.35	.29423570E+07
635.00	89.34	.29423570E+07
640.00	89.33	.29423570E+07
645.00	89.32	.29423570E+07
650.00	89.31	.29423570E+07
655.00	89.31	.29423570E+07
660.00	89.30	.29423570E+07
665.00	89.29	.29423570E+07

Figure 8.12   (Continued)

670.00	89.29	.29423570E+07
675.00	89.28	.29423570E+07
680.00	89.27	.29423570E+07
685.00	89.27	.29423570E+07
690.00	89.26	.29423570E+07
695.00	89.25	.29423570E+07
700.00	89.25	.29423570E+07
705.00	89.24	.29423570E+07
710.00	89.23	.29423570E+07
715.00	89.23	.29423570E+07
720.00	89.22	.29423570E+07
725.00	89.21	.29423570E+07
730.00	89.21	.29423570E+07
735.00	89.20	.29423570E+07
740.00	89.19	.29423570E+07
745.00	89.19	.29423570E+07
750.00	89.18	.29423570E+07
755.00	89.18	.29423570E+07
760.00	89.17	.29423570E+07
765.00	89.16	.29423570E+07
770.00	89.16	.29423570E+07
775.00	89.15	.29423570E+07
780.00	89.15	.29423570E+07
785.00	89.14	.29423570E+07
790.00	89.14	.29423570E+07
795.00	89.13	.29423570E+07
800.00	89.12	29423570E+07

TSKIP =   200.0 HOURS   MAX MODELED TEMPERATURE =   95.22 AT 213.00 HOURS

Figure 8.12   (Continued)

table input, respectively.   The parameter ISPRAY = 1 specifies the regression
spray performance model.   The parameters ISPRON = 0, TSKIP = 0, NITER = 150,
and DTITER = 5 specify that the program should iterate from TSKIP and TSPRON = 0
to 750 hr in 5-hr increments.   The parameter NPRINT = 5000 effectively suppresses
intermediate output so that only the temperature peak for each run is outputted.

The output for this run is shown printed in Figure 8.16 and plotted in Figure 8.17.
From Figure 8.17, there appear to be two temperature peaks, each about 94.0°F.
The first occurs at a value of TSKIP of about 425.0 hr and the second roughly
1 day later at a value of TSKIP of about 447.0 hr.   The value of TSKIP determined
from procedure 1 was 438.0 hr.   For a value of TSKIP = 438.0 hr, the temperature
peak from Figure 8.17 is about 93.7°F.   Therefore, a relatively small error of
about 0.3°F would be made by relying on the estimate on TSKIP from procedure 1.

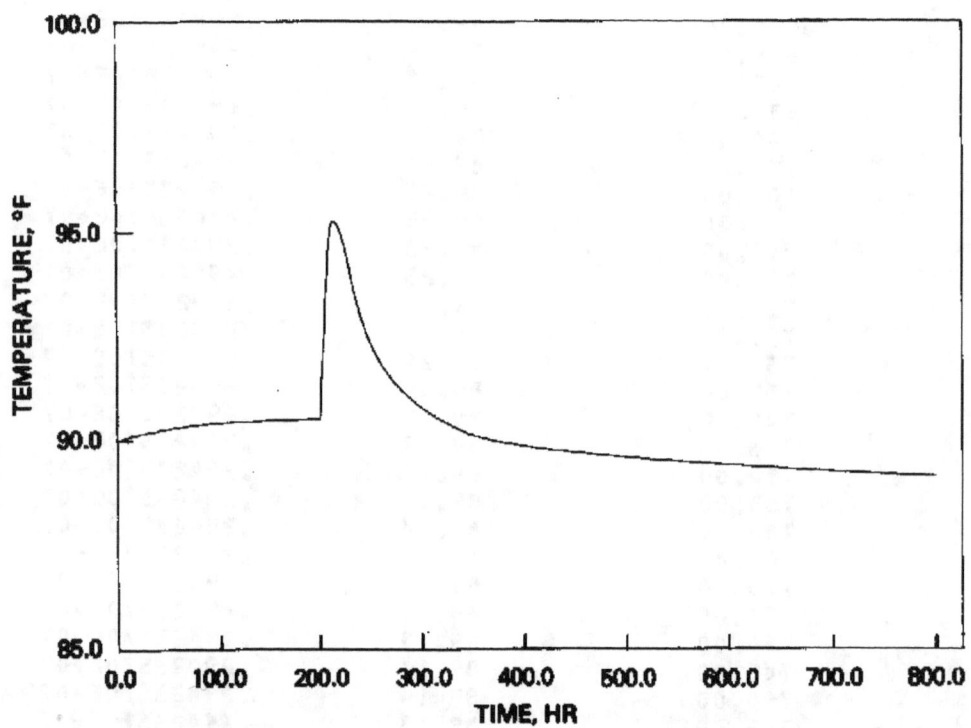

Figure 8.13 Pond temperature in response to design-basis heat load and constant meteorology

The input for the second run of procedure 2 is shown set up in Figure 8.18 for the second peak at TSKIP = 447.0 hr, with ISPRAY = 2, which specifies the rigorous spray model. Output from this run is shown printed in Figure 8.19 and plotted in Figure 8.20. The peak temperature predicted is 93.91°F.

Also shown plotted in Figure 8.20 is the output from the run repeated with the regression spray model (ISPRAY = 1). The agreement between the rigorous and regression ISPRAY performance models is excellent, especially for the highest temperatures. The regression spray model predicts a slightly higher peak temperature of 93.97°F.

```
-.60637276E+00    .40195127E-03    .38449863E-02    .18230236E-02
-.34078270E-01    .30138737E+00  -.25690451E+01    .65576685E-01
-.73791051E-03    .26319278E-05    .35669730E-02    .12911864E-01
-.39275022E-04  -.41450389E-01    .14646531E-03  -.33234415E-03
 .41560445E-03  -.12268707E-02    .11416664E-01  -.86122112E-01
 .28767122E-02  -.29725976E-04    .10168749E-06  -.27394599E-03
 .28406611E-04    .22034012E-05
$PARAM WID=183,ALEN=283,HT=12.0,THETA=71.0,VELO=22.47,R=.095,
   YO=5.0,WDRO=0,NDRIFT=6,DWDR=10,FDRIFT=.0005,.00058,.001914,.004346,
   .00789,.0143305
1176
$HFT NH=14.,TH=0.,.01,1.,1.1,1.9,3.9,5.,8.,12.,24.,29.,140.,840.,2000.,
   HEAT(1)=0.,0.,.85E9,2*.51E9,.5E9,.68E9,.6E9,.4E9,.31E9,.27E9,
   .21E9,.18E9,.1E9,FLOW=14*205200.0$
          COMBINED RUN WITH RIGOROUS MODEL
$INLIST VZERO=2942357,A=422000,NSTEPS=1600,NPRINT=10,Q1=0,F1=1,IEVAP=1,
   DT=.5,IMET=0,ISPRAY=2,
   TSPRON=438,TSKIP=438 $
```

Figure 8.14   Final run of program SPRPND, procedure 1

```
-.60637276E+00    .40195127E-03    .38449863E-02    .18230236E-02
-.34078270E-01    .30138737E+00  -.25690451E+01    .65576685E-01
-.73791051E-03    .26319278E-05    .35669730E-02    .12911864E-01
-.39275022E-04  -.41450389E-01    .14646531E-03  -.33234415E-03
 .41560445E-03  -.12268707E-02    .11416664E-01  -.86122112E-01
 .28767122E-02  -.29725976E-04    .10168749E-06  -.27394599E-03
 .28406611E-04    .22034012E-05
$PARAM WID=183,ALEN=283,HT=12.0,THETA=71.0,VELO=22.47,R=.095,
   YO=5.0,WDRO=0,NDRIFT=6,DWDR=10,FDRIFT=.0005,.00058,.001914,.004346,
   .00789,.0143305
1176
$HFT NH=14.,TH=0.,.01,1.,1.1,1.9,3.9,5.,8.,12.,24.,29.,140.,840.,2000.,
   HEAT(1)=0.,0.,.85E9,2*.51E9,.5E9,.68E9,.6E9,.4E9,.31E9,.27E9,
   .21E9,.18E9,.1E9,FLOW=14*205200.0$
          ITERATE TSKIP FROM 0 TO 750 HOURS STEP 5
$INLIST VZERO=2942357,A=422000,NSTEPS=2000,NPRINT=5000,TZERO=80,IMET=0,
   TSPRON=0,TSKIP=0,DTITER=5,NITER=150,ISPRAY=1,IEVAP=1,Q1=0,F1=1,DT=.5$
```

Figure 8.15   Input data for program SPRPND, iterative run, procedure 2

ITERATE TSKIP FROM 0 TO 750 HOURS STEP 5

SPRAY FIELD PARAMETERS
****
INITIAL VELOCITY OF DROPS LEAVING NOZZLE, VELO =   684.89 CM/SEC
INITIAL ANGLE OF DROPS TO HOR., THETA =   1.239 RADIANS
GEOMETRIC MEAN RADIUS OF DROPS, R =   .0950 CM
HEIGHT OF SPRAY FIELD, HT =   365.76 CM
WIDTH OF SPRAY FIELD, WID =   5577.8 CM
LENGTH OF SPRAY FIELD, ALEN =   8625.8 CM
HEIGHT OF SPRAY NOZZLES ABOVE POND SURFACE, YO =   152.4
HEADING OF WIND W.R.T.LONG AXIS, PHI =   90.00 DEGREES

POND PARAMETERS
****
INITIAL POND VOLUME,VZERO =   2942357.0 CU.FT.
POND SURFACE AREA,A =   422000.0 SQ.FT.
BLOWDOWN AND LEAKAGE,BLOW =   0.00 CU.FT./HR.
NUMBER OF INTEGRATION STEPS,NSTEPS = 2000
PRINT INTERVAL,NPRINT = 5000
INTEGRATION TIMESTEP,DT =   .50 HOURS
INITIAL POND TEMPERATURE,TZERO =   80.00 DEG.F
DELAY FOR HEAT TABLE,TSKIP =   0.00 HRS
BASE HEAT LOAD ADDED TO TABLE,QBASE =   0.00 HRS
BASE FLOW RATE ADDED TO TABLE ,FBASE =   0.   CU.FT./HR.

| HEAT IN | TIME FROM | FLOW IN |
BTU/HR	START	FT**3/HR
0.	0.00	.205E+06
0.	.01	.205E+06
.850E+09	1.00	.205E+06
.510E+09	1.10	.205E+06
.510E+09	1.90	.205E+06
.500E+09	3.90	.205E+06
.680E+09	5.00	.205E+06
.600E+09	8.00	.205E+06
.400E+09	12.00	.205E+06
.310E+09	24.00	.205E+06
.270E+09	29.00	.205E+06
.210E+09	140.00	.205E+06
.180E+09	840.00	.205E+06
.100E+09	2000.00	.205E+06

FOR TIME LESS THAN TSKIP
Q1 =   0.   BTU/HR
F1 =   .100E+01 FT**3/HR

Figure 8.16   Output from program SPRPND, iterative run, procedure 2

METEOROLOGICAL TABLE USED AS INPUT

REGRESSION EQUATIONS USED FOR SPRAY MODEL

SPRAYS WILL BE DELAYED   0.00 HOURS

....

************* MODEL RESULTS *************

```
                        ..TIME.......TEMPERATURE (F)........VOLUME....
                        :  HR                              :   FT**3   :
                        ............................................
TSKIP =    0.0 HOURS    MAX MODELED TEMPERATURE =   91.95 AT  474.50 HOURS
TSKIP =    5.0 HOURS    MAX MODELED TEMPERATURE =   91.94 AT  474.50 HOURS
TSKIP =   10.0 HOURS    MAX MODELED TEMPERATURE =   91.93 AT  474.50 HOURS
TSKIP =   15.0 HOURS    MAX MODELED TEMPERATURE =   91.91 AT  474.50 HOURS
TSKIP =   20.0 HOURS    MAX MODELED TEMPERATURE =   91.90 AT  474.50 HOURS
TSKIP =   25.0 HOURS    MAX MODELED TEMPERATURE =   91.88 AT  474.50 HOURS
TSKIP =   30.0 HOURS    MAX MODELED TEMPERATURE =   91.87 AT  474.50 HOURS
TSKIP =   35.0 HOURS    MAX MODELED TEMPERATURE =   91.86 AT  474.50 HOURS
TSKIP =   40.0 HOURS    MAX MODELED TEMPERATURE =   91.85 AT  474.50 HOURS
TSKIP =   45.0 HOURS    MAX MODELED TEMPERATURE =   91.83 AT  474.50 HOURS
TSKIP =   50.0 HOURS    MAX MODELED TEMPERATURE =   91.82 AT  474.50 HOURS
TSKIP =   55.0 HOURS    MAX MODELED TEMPERATURE =   91.81 AT  474.50 HOURS
TSKIP =   60.0 HOURS    MAX MODELED TEMPERATURE =   91.80 AT  474.50 HOURS
TSKIP =   65.0 HOURS    MAX MODELED TEMPERATURE =   91.80 AT  474.50 HOURS
TSKIP =   70.0 HOURS    MAX MODELED TEMPERATURE =   91.79 AT  474.50 HOURS
TSKIP =   75.0 HOURS    MAX MODELED TEMPERATURE =   91.78 AT  474.50 HOURS
TSKIP =   80.0 HOURS    MAX MODELED TEMPERATURE =   91.77 AT  474.50 HOURS
TSKIP =   85.0 HOURS    MAX MODELED TEMPERATURE =   91.77 AT  474.50 HOURS
TSKIP =   90.0 HOURS    MAX MODELED TEMPERATURE =   91.76 AT  474.50 HOURS
TSKIP =   95.0 HOURS    MAX MODELED TEMPERATURE =   91.76 AT  474.50 HOURS
TSKIP =  100.0 HOURS    MAX MODELED TEMPERATURE =   91.75 AT  474.50 HOURS
TSKIP =  105.0 HOURS    MAX MODELED TEMPERATURE =   91.75 AT  474.50 HOURS
TSKIP =  110.0 HOURS    MAX MODELED TEMPERATURE =   91.74 AT  474.50 HOURS
TSKIP =  115.0 HOURS    MAX MODELED TEMPERATURE =   91.74 AT  474.50 HOURS
TSKIP =  120.0 HOURS    MAX MODELED TEMPERATURE =   91.73 AT  475.00 HOURS
TSKIP =  125.0 HOURS    MAX MODELED TEMPERATURE =   91.73 AT  475.00 HOURS
TSKIP =  130.0 HOURS    MAX MODELED TEMPERATURE =   91.73 AT  475.00 HOURS
TSKIP =  135.0 HOURS    MAX MODELED TEMPERATURE =   91.72 AT  475.00 HOURS
TSKIP =  140.0 HOURS    MAX MODELED TEMPERATURE =   91.72 AT  475.00 HOURS
TSKIP =  145.0 HOURS    MAX MODELED TEMPERATURE =   91.72 AT  475.00 HOURS
TSKIP =  150.0 HOURS    MAX MODELED TEMPERATURE =   91.72 AT  475.00 HOURS
TSKIP =  155.0 HOURS    MAX MODELED TEMPERATURE =   91.72 AT  475.00 HOURS
TSKIP =  160.0 HOURS    MAX MODELED TEMPERATURE =   91.71 AT  475.00 HOURS
TSKIP =  165.0 HOURS    MAX MODELED TEMPERATURE =   91.71 AT  475.00 HOURS
TSKIP =  170.0 HOURS    MAX MODELED TEMPERATURE =   91.71 AT  475.00 HOURS
TSKIP =  175.0 HOURS    MAX MODELED TEMPERATURE =   91.71 AT  475.00 HOURS
TSKIP =  180.0 HOURS    MAX MODELED TEMPERATURE =   91.70 AT  475.00 HOURS
TSKIP =  185.0 HOURS    MAX MODELED TEMPERATURE =   91.70 AT  475.00 HOURS
TSKIP =  190.0 HOURS    MAX MODELED TEMPERATURE =   91.70 AT  475.00 HOURS
TSKIP =  195.0 HOURS    MAX MODELED TEMPERATURE =   91.70 AT  475.00 HOURS
TSKIP =  200.0 HOURS    MAX MODELED TEMPERATURE =   91.70 AT  475.00 HOURS
TSKIP =  205.0 HOURS    MAX MODELED TEMPERATURE =   91.70 AT  475.00 HOURS
```

Figure 8.16   (Continued)

```
TSKIP =    210.0 HOURS    MAX MODELED TEMPERATURE =    91.70 AT    475.00 HOURS
TSKIP =    215.0 HOURS    MAX MODELED TEMPERATURE =    91.70 AT    475.00 HOURS
TSKIP =    220.0 HOURS    MAX MODELED TEMPERATURE =    91.70 AT    475.00 HOURS
TSKIP =    225.0 HOURS    MAX MODELED TEMPERATURE =    91.70 AT    475.00 HOURS
TSKIP =    230.0 HOURS    MAX MODELED TEMPERATURE =    91.70 AT    475.00 HOURS
TSKIP =    235.0 HOURS    MAX MODELED TEMPERATURE =    91.70 AT    475.00 HOURS
TSKIP =    240.0 HOURS    MAX MODELED TEMPERATURE =    91.70 AT    475.00 HOURS
TSKIP =    245.0 HOURS    MAX MODELED TEMPERATURE =    91.71 AT    475.00 HOURS
TSKIP =    250.0 HOURS    MAX MODELED TEMPERATURE =    91.71 AT    475.00 HOURS
TSKIP =    255.0 HOURS    MAX MODELED TEMPERATURE =    91.71 AT    475.00 HOURS
TSKIP =    260.0 HOURS    MAX MODELED TEMPERATURE =    91.71 AT    475.00 HOURS
TSKIP =    265.0 HOURS    MAX MODELED TEMPERATURE =    91.72 AT    475.00 HOURS
TSKIP =    270.0 HOURS    MAX MODELED TEMPERATURE =    91.72 AT    450.50 HOURS
TSKIP =    275.0 HOURS    MAX MODELED TEMPERATURE =    91.74 AT    450.50 HOURS
TSKIP =    280.0 HOURS    MAX MODELED TEMPERATURE =    91.75 AT    450.50 HOURS
TSKIP =    285.0 HOURS    MAX MODELED TEMPERATURE =    91.76 AT    450.50 HOURS
TSKIP =    290.0 HOURS    MAX MODELED TEMPERATURE =    91.77 AT    450.50 HOURS
TSKIP =    295.0 HOURS    MAX MODELED TEMPERATURE =    91.81 AT    401.00 HOURS
TSKIP =    300.0 HOURS    MAX MODELED TEMPERATURE =    91.86 AT    401.00 HOURS
TSKIP =    305.0 HOURS    MAX MODELED TEMPERATURE =    91.92 AT    401.00 HOURS
TSKIP =    310.0 HOURS    MAX MODELED TEMPERATURE =    91.99 AT    401.00 HOURS
TSKIP =    315.0 HOURS    MAX MODELED TEMPERATURE =    92.06 AT    401.00 HOURS
TSKIP =    320.0 HOURS    MAX MODELED TEMPERATURE =    92.14 AT    401.00 HOURS
TSKIP =    325.0 HOURS    MAX MODELED TEMPERATURE =    92.22 AT    401.00 HOURS
TSKIP =    330.0 HOURS    MAX MODELED TEMPERATURE =    92.31 AT    401.00 HOURS
TSKIP =    335.0 HOURS    MAX MODELED TEMPERATURE =    92.41 AT    401.00 HOURS
TSKIP =    340.0 HOURS    MAX MODELED TEMPERATURE =    92.52 AT    401.00 HOURS
TSKIP =    345.0 HOURS    MAX MODELED TEMPERATURE =    92.65 AT    401.00 HOURS
TSKIP =    350.0 HOURS    MAX MODELED TEMPERATURE =    92.76 AT    401.00 HOURS
TSKIP =    355.0 HOURS    MAX MODELED TEMPERATURE =    92.88 AT    401.00 HOURS
TSKIP =    360.0 HOURS    MAX MODELED TEMPERATURE =    93.02 AT    401.00 HOURS
TSKIP =    365.0 HOURS    MAX MODELED TEMPERATURE =    93.19 AT    401.00 HOURS
TSKIP =    370.0 HOURS    MAX MODELED TEMPERATURE =    93.31 AT    401.00 HOURS
TSKIP =    375.0 HOURS    MAX MODELED TEMPERATURE =    93.26 AT    401.50 HOURS
TSKIP =    380.0 HOURS    MAX MODELED TEMPERATURE =    92.88 AT    402.00 HOURS
TSKIP =    385.0 HOURS    MAX MODELED TEMPERATURE =    92.82 AT    450.50 HOURS
TSKIP =    390.0 HOURS    MAX MODELED TEMPERATURE =    92.93 AT    450.50 HOURS
TSKIP =    395.0 HOURS    MAX MODELED TEMPERATURE =    93.04 AT    450.50 HOURS
TSKIP =    400.0 HOURS    MAX MODELED TEMPERATURE =    93.16 AT    450.50 HOURS
TSKIP =    405.0 HOURS    MAX MODELED TEMPERATURE =    93.29 AT    450.50 HOURS
TSKIP =    410.0 HOURS    MAX MODELED TEMPERATURE =    93.48 AT    450.00 HOURS
TSKIP =    415.0 HOURS    MAX MODELED TEMPERATURE =    93.72 AT    450.00 HOURS
TSKIP =    420.0 HOURS    MAX MODELED TEMPERATURE =    93.90 AT    450.00 HOURS
TSKIP =    425.0 HOURS    MAX MODELED TEMPERATURE =    94.00 AT    450.50 HOURS
TSKIP =    430.0 HOURS    MAX MODELED TEMPERATURE =    93.80 AT    451.50 HOURS
TSKIP =    435.0 HOURS    MAX MODELED TEMPERATURE =    93.52 AT    475.00 HOURS
TSKIP =    440.0 HOURS    MAX MODELED TEMPERATURE =    93.76 AT    475.00 HOURS
TSKIP =    445.0 HOURS    MAX MODELED TEMPERATURE =    93.92 AT    475.00 HOURS
TSKIP =    450.0 HOURS    MAX MODELED TEMPERATURE =    93.98 AT    475.00 HOURS
TSKIP =    455.0 HOURS    MAX MODELED TEMPERATURE =    93.84 AT    475.50 HOURS
TSKIP =    460.0 HOURS    MAX MODELED TEMPERATURE =    93.58 AT    476.50 HOURS
TSKIP =    465.0 HOURS    MAX MODELED TEMPERATURE =    92.95 AT    495.00 HOURS
TSKIP =    470.0 HOURS    MAX MODELED TEMPERATURE =    93.18 AT    497.00 HOURS
TSKIP =    475.0 HOURS    MAX MODELED TEMPERATURE =    93.20 AT    497.50 HOURS
TSKIP =    480.0 HOURS    MAX MODELED TEMPERATURE =    93.14 AT    498.00 HOURS
TSKIP =    485.0 HOURS    MAX MODELED TEMPERATURE =    92.92 AT    499.00 HOURS
TSKIP =    490.0 HOURS    MAX MODELED TEMPERATURE =    91.96 AT    500.50 HOURS
TSKIP =    495.0 HOURS    MAX MODELED TEMPERATURE =    91.41 AT    518.00 HOURS
TSKIP =    500.0 HOURS    MAX MODELED TEMPERATURE =    91.23 AT    518.50 HOURS
TSKIP =    505.0 HOURS    MAX MODELED TEMPERATURE =    90.94 AT    519.00 HOURS
TSKIP =    510.0 HOURS    MAX MODELED TEMPERATURE =    90.41 AT    522.00 HOURS
```

Figure 8.16   (Continued)

```
TSKIP =    515.0 HOURS     MAX MODELED TEMPERATURE =    89.62 AT    526.50 HOURS
TSKIP =    520.0 HOURS     MAX MODELED TEMPERATURE =    89.22 AT    539.50 HOURS
TSKIP =    525.0 HOURS     MAX MODELED TEMPERATURE =    89.05 AT    542.00 HOURS
TSKIP =    530.0 HOURS     MAX MODELED TEMPERATURE =    88.84 AT    544.00 HOURS
TSKIP =    535.0 HOURS     MAX MODELED TEMPERATURE =    88.35 AT    545.50 HOURS
TSKIP =    540.0 HOURS     MAX MODELED TEMPERATURE =    86.79 AT    548.00 HOURS
TSKIP =    545.0 HOURS     MAX MODELED TEMPERATURE =    85.86 AT    762.00 HOURS
TSKIP =    550.0 HOURS     MAX MODELED TEMPERATURE =    85.86 AT    762.00 HOURS
TSKIP =    555.0 HOURS     MAX MODELED TEMPERATURE =    85.87 AT    762.00 HOURS
TSKIP =    560.0 HOURS     MAX MODELED TEMPERATURE =    85.87 AT    762.00 HOURS
TSKIP =    565.0 HOURS     MAX MODELED TEMPERATURE =    85.88 AT    762.00 HOURS
TSKIP =    570.0 HOURS     MAX MODELED TEMPERATURE =    85.89 AT    762.00 HOURS
TSKIP =    575.0 HOURS     MAX MODELED TEMPERATURE =    85.90 AT    762.00 HOURS
TSKIP =    580.0 HOURS     MAX MODELED TEMPERATURE =    85.91 AT    762.00 HOURS
TSKIP =    585.0 HOURS     MAX MODELED TEMPERATURE =    85.92 AT    762.00 HOURS
TSKIP =    590.0 HOURS     MAX MODELED TEMPERATURE =    85.94 AT    762.00 HOURS
TSKIP =    595.0 HOURS     MAX MODELED TEMPERATURE =    85.96 AT    762.00 HOURS
TSKIP =    600.0 HOURS     MAX MODELED TEMPERATURE =    85.98 AT    762.00 HOURS
TSKIP =    605.0 HOURS     MAX MODELED TEMPERATURE =    86.00 AT    762.00 HOURS
TSKIP =    610.0 HOURS     MAX MODELED TEMPERATURE =    86.03 AT    762.00 HOURS
TSKIP =    615.0 HOURS     MAX MODELED TEMPERATURE =    86.06 AT    762.00 HOURS
TSKIP =    620.0 HOURS     MAX MODELED TEMPERATURE =    86.10 AT    762.00 HOURS
TSKIP =    625.0 HOURS     MAX MODELED TEMPERATURE =    86.15 AT    762.00 HOURS
TSKIP =    630.0 HOURS     MAX MODELED TEMPERATURE =    86.20 AT    762.00 HOURS
TSKIP =    635.0 HOURS     MAX MODELED TEMPERATURE =    86.25 AT    762.00 HOURS
TSKIP =    640.0 HOURS     MAX MODELED TEMPERATURE =    86.31 AT    762.00 HOURS
TSKIP =    645.0 HOURS     MAX MODELED TEMPERATURE =    86.37 AT    762.00 HOURS
TSKIP =    650.0 HOURS     MAX MODELED TEMPERATURE =    86.44 AT    762.00 HOURS
TSKIP =    655.0 HOURS     MAX MODELED TEMPERATURE =    86.50 AT    762.00 HOURS
TSKIP =    660.0 HOURS     MAX MODELED TEMPERATURE =    86.57 AT    762.00 HOURS
TSKIP =    665.0 HOURS     MAX MODELED TEMPERATURE =    86.65 AT    762.00 HOURS
TSKIP =    670.0 HOURS     MAX MODELED TEMPERATURE =    86.73 AT    762.00 HOURS
TSKIP =    675.0 HOURS     MAX MODELED TEMPERATURE =    86.81 AT    762.00 HOURS
TSKIP =    680.0 HOURS     MAX MODELED TEMPERATURE =    86.90 AT    762.00 HOURS
TSKIP =    685.0 HOURS     MAX MODELED TEMPERATURE =    86.99 AT    762.00 HOURS
TSKIP =    690.0 HOURS     MAX MODELED TEMPERATURE =    87.09 AT    762.00 HOURS
TSKIP =    695.0 HOURS     MAX MODELED TEMPERATURE =    87.19 AT    762.00 HOURS
TSKIP =    700.0 HOURS     MAX MODELED TEMPERATURE =    87.30 AT    762.00 HOURS
TSKIP =    705.0 HOURS     MAX MODELED TEMPERATURE =    87.41 AT    762.00 HOURS
TSKIP =    710.0 HOURS     MAX MODELED TEMPERATURE =    87.50 AT    762.00 HOURS
TSKIP =    715.0 HOURS     MAX MODELED TEMPERATURE =    87.58 AT    762.00 HOURS
TSKIP =    720.0 HOURS     MAX MODELED TEMPERATURE =    87.68 AT    762.00 HOURS
TSKIP =    725.0 HOURS     MAX MODELED TEMPERATURE =    87.80 AT    762.00 HOURS
TSKIP =    730.0 HOURS     MAX MODELED TEMPERATURE =    87.85 AT    762.00 HOURS
TSKIP =    735.0 HOURS     MAX MODELED TEMPERATURE =    87.80 AT    762.50 HOURS
TSKIP =    740.0 HOURS     MAX MODELED TEMPERATURE =    87.51 AT    764.50 HOURS
TSKIP =    745.0 HOURS     MAX MODELED TEMPERATURE =    87.04 AT    767.00 HOURS


TSKIP =    750.0 HOURS     MAX MODELED TEMPERATURE =    86.51 AT    770.00 HOURS
```

Figure 8.16   (Continued)

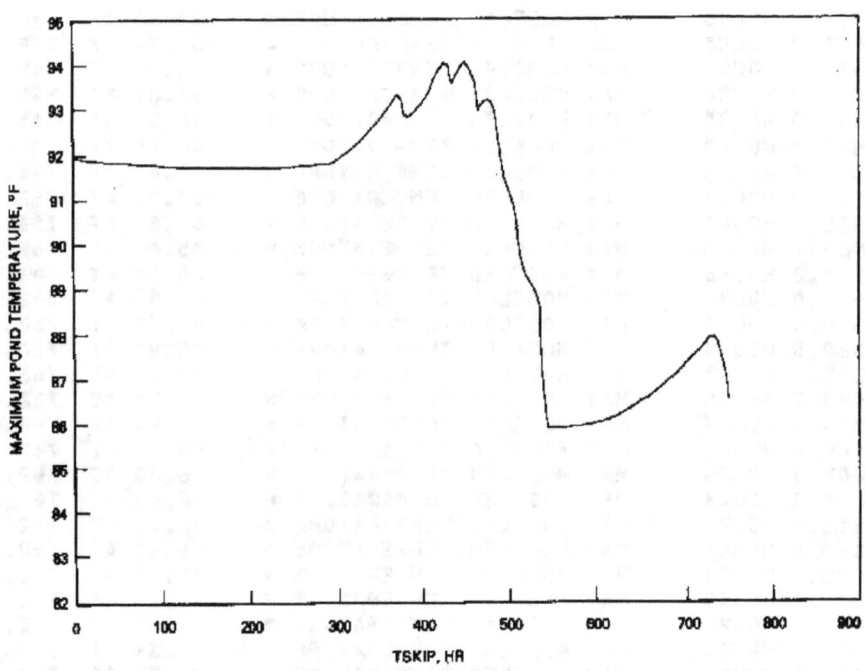

Figure 8.17 Effect of starting time for design-basis accident on peak pond temperature

8.7 Correction Factors for Geographic Differences Between Site and Meteorological Station--Program COMET2

Program COMET2 is used to estimate the differences in the meteorological data bases of the site and the point at which the long-term meteorological data were taken. Monthly average values of wet-bulb temperature, dry-bulb temperature, rms windspeed, barometric pressure, dewpoint temperature, and solar radiation were obtained from program SPSCAN for a 15-month period corresponding to the period of onsite data availability at the Susquehanna site. The input data for COMET2 are shown in Figure 8.21. The output is shown printed in Figure 8.22 and plotted in Figures 8.23 and 8.24. It is clear from the output that there are biases in the two data sets. The average bias for the Susquehanna-site data indicates that the spray-pond temperature should be about 1.36°F lower than predicted from program SPRPND.* The evaporation should also be less by

---

*Although the points in Figure 8.23 fall on both sides of the 45° diagonal line, the Harrisburg data are most conservative at the higher temperatures, which is the region of greater concern.

163,464 ft^3. The Harrisburg data are, therefore, conservative. Although the peak temperature and evaporation could have been corrected by the above amounts, it is suggested that the corrections be performed only if they lead to greater conservatism.

```
-.60637276E+00    .40195127E-03    .38449863E-02    .18230236E-02
-.34078270E-01    .30138737E+00   -.25690451E+01    .65576685E-01
-.73791051E-03    .26319278E-05    .35669730E-02    .12911864E-01
-.39275022E-04   -.41450389E-01    .14646531E-03   -.33234415E-03
 .41560445E-03   -.12268707E-02    .11416664E-01   -.86122112E-01
 .28767122E-02   -.29725976E-04    .10168749E-06   -.27394599E-03
 .28406611E-04    .22034012E-05
$PARAM WID=183,ALEN=283,HT=12.0,THETA=71.0,VELO=22.47,R=.095,
   Y0=5.0,WDRO=0,NDRIFT=6,DWDR=10,FDRIFT=.0005,.00058,.001914,.004346,
   .00789,.014330$
1176
$HFT NH=14.,TH=0.,.01,1.,1.1,1.9,3.9,5.,8.,12.,24.,29.,140.,840.,2000.,
   HEAT(1)=0.,0.,.85E9,2*.51E9,.5E9,.68E9,.6E9,.4E9,.31E9,.27E9,
   .21E9,.18E9,.1E9,FLOW=14*205200.0$
         COMBINED RUN WITH RIGOROUS MODEL
$INLIST VZERO=2942357,A=422000,NSTEPS=1600,NPRINT=10,Q1=0,F1=1,IEVAP=1,
   DT=.5,IMET=0,ISPRAY=2,
   TSPRON=447,TSKIP=447$
         COMBINED RUN WITH REGRESSION MODEL
$INLIST VZERO=2942357,A=422000,NSTEPS=1600,NPRINT=10,Q1=0,F1=1,IEVAP=1,
   DT=.5,IMET=0,ISPRAY=1,
   TSPRON=447,TSKIP=447$
```

Figure 8.18   Input for program SPRPND, combined runs for rigorous and regression spray models, TSKIP = 447.0 hours

8.8 Statistical Adjustments

Program SPSCAN calculates the yearly maximum temperature and 30-day water loss for each year of record. The maximum likelihood and 5% and 95% confidence limits are generated for these data. Using the procedures outlined in Appendix A, it is possible to construct the plots of temperature and evaporation, respectively, versus recurrence interval shown in Figures 8.25 and 8.26. It is then possible to estimate correction factors based on the recurrence intervals of the peak temperature and evaporation found. If, for example, the 100-yr recurrence-interval meteorology were chosen as the basis for the temperature and evaporation conditions, correction factors could be developed for the final answer as demonstrated below:

COMBINED RUN WITH RIGOROUS MODEL

SPRAY FIELD PARAMETERS
****
INITIAL VELOCITY OF DROPS LEAVING NOZZLE, VELO =   684.89 CM/SEC
INITIAL ANGLE OF DROPS TO HOR., THETA =   1.239 RADIANS
GEOMETRIC MEAN RADIUS OF DROPS, R =   .0950 CM
HEIGHT OF SPRAY FIELD, HT =   365.76 CM
WIDTH OF SPRAY FIELD, WID =   5577.8 CM
LENGTH OF SPRAY FIELD, ALEN =   8625.8 CM
HEIGHT OF SPRAY NOZZLES ABOVE POND SURFACE, YO =   152.4
HEADING OF WIND W.R.T.LONG AXIS, PHI =   90.00 DEGREES

POND PARAMETERS
****
INITIAL POND VOLUME,VZERO =   2942357.0 CU.FT.
POND SURFACE AREA,A =   422000.0 SQ.FT.
BLOWDOWN AND LEAKAGE,BLOW =   0.00 CU.FT./HR.
NUMBER OF INTEGRATION STEPS,NSTEPS =   1600
PRINT INTERVAL,NPRINT =   10
INTEGRATION TIMESTEP,DT =   .50 HOURS
INITIAL POND TEMPERATURE,TZERO =   80.00 DEG.F
DELAY FOR HEAT TABLE,TSKIP =   447.00 HRS
BASE HEAT LOAD ADDED TO TABLE,QBASE =   0.00 HRS
BASE FLOW RATE ADDED TO TABLE ,FBASE =   0.   CU.FT./HR.

| HEAT IN | TIME FROM | FLOW IN |
BTU/HR	START	FT**3/HR
0.	0.00	.205E+06
0.	.01	.205E+06
.850E+09	1.00	.205E+06
.510E+09	1.10	.205E+06
.510E+09	1.90	.205E+06
.500E+09	3.90	.205E+06
.680E+09	5.00	.205E+06
.600E+09	8.00	.205E+06
.400E+09	12.00	.205E+06
.310E+09	24.00	.205E+06
.270E+09	29.00	.205E+06
.210E+09	140.00	.205E+06
.180E+09	840.00	.205E+06
.100E+09	2000.00	.205E+06

Figure 8.19   Output from program SPRPND, run for rigorous spray model, TSKIP = 447.0 hours

```
          FOR TIME LESS THAN TSKIP
          Q1 =     0.         BTU/HR
          F1 =      .100E+01 FT**3/HR

METEOROLOGICAL TABLE USED AS INPUT


RIGOROUS SPRAY MODEL CHOSEN


SPRAYS WILL BE DELAYED    447.00 HOURS


            ...................................


            ************* MODEL RESULTS *************


       ..TIME........TEMPERATURE (F)........VOLUME....
       :  HR                              :   FT**3   :
       ............................................
           5.00              79.03          .29423570E+07
          10.00              79.24          .29423570E+07
          15.00              79.90          .29423570E+07
          20.00              79.25          .29423570E+07
          25.00              77.87          .29423570E+07
          30.00              76.33          .29423570E+07
          35.00              75.98          .29423570E+07
          40.00              76.19          .29423570E+07
          45.00              75.66          .29423570E+07
          50.00              74.64          .29423570E+07
          55.00              73.33          .29423570E+07
          60.00              72.04          .29423570E+07
          65.00              71.28          .29423570E+07
          70.00              70.24          .29423570E+07
          75.00              69.24          .29423570E+07
          80.00              68.50          .29423570E+07
          85.00              68.33          .29423570E+07
          90.00              68.47          .29423570E+07
          95.00              67.96          .29423570E+07
         100.00              67.25          .29423570E+07
         105.00              66.90          .29423570E+07
         110.00              67.90          .29423570E+07
         115.00              68.26          .29423570E+07
         120.00              67.87          .29423570E+07
         125.00              67.24          .29423570E+07
         130.00              67.56          .29423570E+07
         135.00              68.31          .29423570E+07
         140.00              68.22          .29423570E+07
         145.00              67.67          .29423570E+07
         150.00              67.04          .29423570E+07
         155.00              67.00          .29423570E+07
```

Figure 8.19   (Continued)

160.00	67.83	.29423570E+07
165.00	67.93	.29423570E+07
170.00	67.72	.29423570E+07
175.00	67.67	.29423570E+07
180.00	69.07	.29423570E+07
185.00	69.96	.29423570E+07
190.00	69.88	.29423570E+07
195.00	69.58	.29423570E+07
200.00	69.78	.29423570E+07
205.00	71.45	.29423570E+07
210.00	72.66	.29423570E+07
215.00	72.57	.29423570E+07
220.00	72.25	.29423570E+07
225.00	72.57	.29423570E+07
230.00	73.59	.29423570E+07
235.00	73.87	.29423570E+07
240.00	73.55	.29423570E+07
245.00	73.13	.29423570E+07
250.00	73.11	.29423570E+07
255.00	74.03	.29423570E+07
260.00	73.91	.29423570E+07
265.00	73.13	.29423570E+07
270.00	72.59	.29423570E+07
275.00	73.59	.29423570E+07
280.00	74.80	.29423570E+07
285.00	74.77	.29423570E+07
290.00	74.32	.29423570E+07
295.00	73.97	.29423570E+07
300.00	75.25	.29423570E+07
305.00	76.61	.29423570E+07
310.00	76.40	.29423570E+07
315.00	75.83	.29423570E+07
320.00	75.78	.29423570E+07
325.00	77.12	.29423570E+07
330.00	77.57	.29423570E+07
335.00	77.10	.29423570E+07
340.00	76.60	.29423570E+07
345.00	76.32	.29423570E+07
350.00	77.35	.29423570E+07
355.00	77.53	.29423570E+07
360.00	77.12	.29423570E+07
365.00	76.60	.29423570E+07
370.00	76.74	.29423570E+07
375.00	77.73	.29423570E+07
380.00	78.00	.29423570E+07
385.00	77.67	.29423570E+07
390.00	77.29	.29423570E+07
395.00	78.06	.29423570E+07
400.00	79.55	.29423570E+07
405.00	79.75	.29423570E+07
410.00	79.38	.29423570E+07
415.00	79.04	.29423570E+07
420.00	80.17	.29423570E+07
425.00	81.19	.29423570E+07

Figure 8.19   (Continued)

430.00	81.04	.29423570E+07
435.00	80.53	.29423570E+07
440.00	80.38	.29423570E+07
445.00	81.42	.29423570E+07
450.00	85.34	.29272932E+07
455.00	89.55	.28972500E+07
460.00	90.63	.28696068E+07
465.00	90.96	.28455993E+07
470.00	92.43	.28229635E+07
475.00	93.90	.28039529E+07
480.00	92.92	.27846809E+07
485.00	91.33	.27653170E+07
490.00	91.46	.27467519E+07
495.00	92.25	.27262648E+07
500.00	91.93	.27065416E+07
505.00	90.13	.26870599E+07
510.00	88.93	.26694924E+07
515.00	89.19	.26522187E+07
520.00	89.40	.26325687E+07
525.00	88.34	.26131694E+07
530.00	87.37	.25969744E+07
535.00	86.76	.25816960E+07
540.00	86.74	.25629735E+07
545.00	86.37	.25429982E+07
550.00	84.18	.25224251E+07
555.00	81.35	.25029815E+07
560.00	79.72	.24860614E+07
565.00	79.61	.24673550E+07
570.00	79.31	.24491142E+07
575.00	79.12	.24352591E+07
580.00	79.10	.24227987E+07
585.00	79.35	.24102278E+07
590.00	80.05	.23967512E+07
595.00	80.42	.23840885E+07
600.00	80.27	.23718274E+07
605.00	79.62	.23593050E+07
610.00	80.23	.23470600E+07
615.00	81.39	.23322636E+07
620.00	80.56	.23156647E+07
625.00	79.23	.23021190E+07
630.00	77.37	.22885307E+07
635.00	77.42	.22755886E+07
640.00	79.20	.22632377E+07
645.00	79.84	.22507591E+07
650.00	79.44	.22381073E+07
655.00	78.64	.22256378E+07
660.00	79.15	.22129448E+07
665.00	79.79	.22000894E+07
670.00	79.97	.21880215E+07
675.00	79.24	.21752503E+07
680.00	79.06	.21636144E+07
685.00	79.71	.21524136E+07
690.00	80.65	.21418406E+07
695.00	80.99	.21311774E+07
700.00	80.25	.21192703E+07

Figure 8.19   (Continued)

```
705.00              80.12              .21075767E+07
710.00              82.22              .20960858E+07
715.00              83.20              .20830286E+07
720.00              82.83              .20703926E+07
725.00              81.96              .20579820E+07
730.00              82.92              .20464775E+07
735.00              84.40              .20331592E+07
740.00              84.11              .20187462E+07
745.00              83.14              .20055908E+07
750.00              82.40              .19935346E+07
755.00              83.11              .19821931E+07
760.00              85.22              .19714001E+07
765.00              85.41              .19583762E+07
770.00              84.54              .19452174E+07
775.00              83.38              .19313785E+07
780.00              82.38              .19153827E+07
785.00              80.04              .18978776E+07
790.00              77.97              .18825298E+07
795.00              76.14              .18692831E+07
800.00              74.86              .18563833E+07
```

TSKIP = 447.0 HOURS   MAX MODELED TEMPERATURE =   93.91 AT   475.00 HOURS

Figure 8.19   (Continued)

$\Delta T = T(100\text{-yr recurrence}) - T_{max} = 93.14°F - 92.74°F = +0.40°F$

and

$\Delta EVAP = EVAP(100\text{-yr recurrence}) - EVAP_{max}$
$= 2.435 \times 10^6 - 2.463 \times 10^6 = -28,000 \text{ ft}^3$

The correction factor for the 100-yr recurrence interval is positive for temperatures and, therefore, should be added to the final result from program SPRPND:

Design-basis maximum temperature = 93.91 + 0.40 $\simeq$ 94.3°F

The correction for evaporation is negative and, therefore, should not be added to the results:

Design-basis 30-day evaporation = $2.46 \times 10^6 \text{ ft}^3$

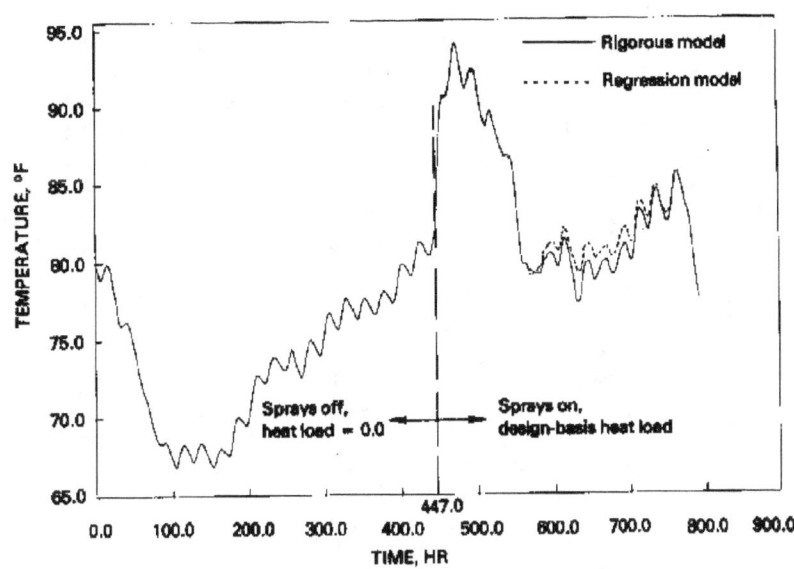

Figure 8.20   Pond temperature in response to ambient meteorology
and design-basis heat load, TSKIP = 447.0 hours

8.9 Conclusion

The maximum pond temperature is predicted to be 94.3°F. The maximum 30-day
evaporation is predicted to be 2.46 x 10⁶ ft³. These results are conservative
because it has been demonstrated that the evaporation and temperature using
Susquehanna-site data would be lower than the results using Harrisburg data.
In addition, the peak evaporation appears to use meteorological data with greater
than a 100-year-recurrence interval.

Water loss from seepage and blowdown or other uses must, of course, be added
to the evaporation and drift losses predicted.

It has been demonstrated that procedure 1 may be used to determine the starting
time for the final calculations with only a small error in peak temperature,
and that the regression spray model is a reliable predictor of the spray
performance determined from the rigorous spray model.

```
-.60637276E+00    .40195127E-03    .38449863E-02    .18230236E-02
-.34078270E-01    .30138737E+00   -.25690451E+01    .65576685E-01
-.73791051E-03    .26319278E-05    .35669730E-02    .12911864E-01
-.39275022E-04   -.41450389E-01    .14646531E-03   -.33234415E-03
 .41560445E-03   -.12268707E-02    .11416664E-01   -.86122112E-01
 .28767122E-02   -.29725976E-04    .10168749E-06   -.27394599E-03
 .28406611E-04    .22034012E-05
$INLIST V=2942357.,A=422000.,HEAT=2.3E8,NDRIFT=6,WDRO=0,DWDR=10.,
    QSPRAY=57,         FDRIFT=.0005,.00058,.001914,.004346,.007890,.01035$
15
52.03    57.38    8.91    1378.4    29.54    49.9    53.8    5.48    0.    29.54
66.35    72.79    6.12    1662.2    29.64    59.1    67.5    3.9     0.    29.64
68.19    76.14    6.43    1884.9    29.63    59.7    67.8    3.7     0.    29.63
68.37    75.38    5.90    1539.1    29.67    60.7    69.4    3.2     0.    29.67
60.89    67.87    7.37    1291.5    29.71    58.    60.4    4.      0.    29.71
54.71    63.47    8.61    1648.3    29.58    51.6    56.7    5.6     0.    29.58
62.61    70.6     7.59    1686.6    29.6     55.9    63.1    4.8     0.    29.60
66.46    72.27    7.54    1763.9    29.64    59.1    68.4    4.12    0.    29.64
68.34    76.47    5.74    1377.3    29.71    59.5    68.    3.39    0.    29.71
59.29    64.24    7.46    1182.6    29.7     53.6    58.5    4.18    0.    29.70
59.57    64.74    6.65    1599.6    29.61    54.6    61.    4.36    0.    29.61
65.36    70.57    7.61    1636.9    29.67    58.    65.7    4.53    0.    29.67
69.36    75.01    6.84    1746.9    29.65    60.3    69.4    3.54    0.    29.65
69.63    75.12    6.75    1507.4    29.7     59.6    68.2    3.94    0.    29.70
58.99    62.82    7.31    1157.    29.76    53.2    57.9    4.47    0.    29.76
```

Figure 8.21   Input data for program COMET2

DIFFERENCES IN STEADY STATE TEMPERATURES AND WATER USE FOR SUBJECT SPRAY POND
USING MONTHLY AVERAGE VALUES OF WET BULB,DRY BULB,WIND SPEED,AND SOLAR RADIATION FROM UNSITE
AND OFFSITE MET STATIONS

```
TIMESTEP IN ITERATION DTIME =          12.527 HOURS
VOLUME OF POND, V =      2942357.0 FT**3
SURFACE AREA OF POND, A =       422000.0 FT**2
RATE OF SPRAYING, QSPRAY =           57.0 FT**3/SEC
STEADY HEAT LOAD, HEAT = 230000000.0 BTU/HR
LOWER LIMIT OF WIND IN DRIFT TABLE WDR0 =         0.00 MPH
INCREMENT IN DRIFT TABLE,DWDR =       10.00 MPH
```

DRIFT LOSS TABLE
WIND SPEED, MPH   DRIFT LOSS FRACTION

```
         0.00         .00050000
        10.00         .00058000
        20.00         .00191400
        30.00         .00434600
        40.00         .00759000
        50.00         .01433000
```

WET BULB SOLAR RAD. (DEG. F)	DRY BULB (DEG. F)	WIND SPEED (MPH)	(BTU/FT**2/DY)	PB INCHES HG	POND TEMP (DEG. F)	EVAPORATION FT**3
DATA SET 1 52.03	57.38	8.91	1378.40	29.54	72.43	2043639.16
DATA SET 2 49.90	53.80	5.48	1378.40	29.54	77.65	1952527.68
			E2-E1 = 5.221		EVAP2-EVAP1 =	-91111.5

DIFFERENCES IN E BETWEEN DATA SET 2 AND DATA SET 1 BY PARAMETER

```
DIFFERENCE DUE TO WET BULB =              -1.098 DEG. F
DIFFERENCE DUE TO DRY BULB TEMP. =         -.183 DEG. F
DIFFERENCE DUE TO WIND SPEED =             5.367 DEG. F
DIFFERENCE DUE TO INSOLATION =             0.000 DEG. F
DIFFERENCE DUE TO BAROMETRIC PRESSURE =    0.000 DEG. F
SUMMATION OF INDIVIDUAL DIFFERENCES =      4.105 DEG. F
```

Figure 8.22   Input deck for program SPRCO

```
*************************************************************************************

            WET BULB   DRY BULB   WIND SPEED
            SOLAR RAD.   (DEG,F)    (MPH)    (BTU/FT**2/DY) INCHES HG   POND TEMP   EVAPORATION
            (DEG. F)                                           PB      (DEG. F)      FT**3

DATA SET 1   66.35      72.79      6.12      1662.20       29.44       83.33      2292962.49

DATA SET 2   59.10      67.50      3.90      1662.20       29.64       80.46      2180291.13

                                           E2-E1 = -2.876        EVAP2-EVAP1 = -112671.4
```

DIFFERENCES IN E BETWEEN DATA SET 2 AND DATA SET 1 BY PARAMETER

```
DIFFERENCE DUE TO WET BULB =            -3.602 DEG. F
DIFFERENCE DUE TO DRY BULB TEMP. =       -.175 DEG. F
DIFFERENCE DUE TO WIND SPEED =            .827 DEG. F
DIFFERENCE DUE TO INSOLATION =           0.000 DEG. F
DIFFERENCE DUE TO BAROMETRIC PRESSURE =  0.000 DEG. F
SUMMATION OF INDIVIDUAL DIFFERENCES =   -2.950 DEG. F
```

```
*************************************************************************************

            WET BULB   DRY BULB   WIND SPEED
            SOLAR RAD.   (DEG,F)    (MPH)    (BTU/FT**2/DY) INCHES HG   POND TEMP   EVAPORATION
            (DEG. F)                                           PB      (DEG. F)      FT**3

DATA SET 1   68.19      76.14      6.43      1884.90       29.63       84.85      2424193.83

DATA SET 2   59.70      67.80      3.70      1884.90       29.63       81.17      2214258.36

                                           E2-E1 = -3.686        EVAP2-EVAP1 = -209935.5
```

DIFFERENCES IN E BETWEEN DATA SET 2 AND DATA SET 1 BY PARAMETER

```
DIFFERENCE DUE TO WET BULB =            -4.269 DEG. F
DIFFERENCE DUE TO DRY BULB TEMP. =       -.472 DEG. F
DIFFERENCE DUE TO WIND SPEED =           1.023 DEG. F
DIFFERENCE DUE TO INSOLATION =           0.000 DEG. F
DIFFERENCE DUE TO BAROMETRIC PRESSURE =  0.000 DEG. F
SUMMATION OF INDIVIDUAL DIFFERENCES =   -3.717 DEG. F
```

Figure 8.22   (Continued)

```
****************************************************************************

              WET BULB      DRY BULB    WIND SPEED                    PB          POND TEMP     EVAPORATION
              SOLAR RAD.    (DEG.F)     (MPH)      (BTU/FT**2/DY)  INCHES HG     (DEG. F)       FT**3
              (DEG. F)

DATA SET 1    68.37         75.38       5.90        1539.10         29.67         84.55         2342443.19

DATA SET 2    60.70         69.40       3.20        1539.10         29.67         81.30         2206192.05

                                                                 E2-E1 = -3.250               EVAP2-EVAP1 = -136651.1
```

DIFFERENCES IN E BETWEEN DATA SET 2 AND DATA SET 1 BY PARAMETER

```
DIFFERENCE DUE TO WET BULB =                        -3.912 DEG. F
DIFFERENCE DUE TO DRY BULB TEMP. =                   -.382 DEG. F
DIFFERENCE DUE TO WIND SPEED =                        .999 DEG. F
DIFFERENCE DUE TO INSOLATION =                       0.000 DEG. F
DIFFERENCE DUE TO BAROMETRIC PRESSURE =              0.000 DEG. F
SUMMATION OF INDIVIDUAL DIFFERENCES =               -3.295 DEG. F
```

```
****************************************************************************

              WET BULB      DRY BULB    WIND SPEED                    PB          POND TEMP     EVAPORATION
              SOLAR RAD.    (DEG.F)     (MPH)      (BTU/FT**2/DY)  INCHES HG     (DEG. F)       FT**3
              (DEG. F)

DATA SET 1    60.89         67.87       7.37        1291.50         29.71         79.35         2127673.35

DATA SET 2    58.00         60.40       4.00        1291.50         29.71         79.99         1841507.91

                                                                 E2-E1 = .639                EVAP2-EVAP1 = -286165.4
```

DIFFERENCES IN E BETWEEN DATA SET 2 AND DATA SET 1 BY PARAMETER

```
DIFFERENCE DUE TO WET BULB =                        -1.378 DEG. F
DIFFERENCE DUE TO DRY BULB TEMP. =                   -.061 DEG. F
DIFFERENCE DUE TO WIND SPEED =                       1.290 DEG. F
DIFFERENCE DUE TO INSOLATION =                       0.000 DEG. F
DIFFERENCE DUE TO BAROMETRIC PRESSURE =              0.000 DEG. F
SUMMATION OF INDIVIDUAL DIFFERENCES =                -.149 DEG. F
```

Figure 8.22 (Continued)

```
******************************************************************************************

          WET BULB      DRY BULB   WIND SPEED                           POND TEMP    EVAPORATION
          SOLAR RAD.    (DEG,F)      (MPH)    (BTU/FT**2/DY)  PB         (DEG. F)       FT**3
          (DEG. F)                                          INCHES HG

DATA SET 1   54.71       65.47       8.61       1648.80      29.58        75.06      2216888.45

DATA SET 2   51.60       56.70       5.60       1648.80      29.58        78.21      1865384.62

                                               E2-E1 =  3.203      EVAP2-EVAP1 =  -351503.8
```

DIFFERENCES IN E BETWEEN DATA SET 2 AND DATA SET 1 BY PARAMETER

```
DIFFERENCE DUE TO WET BULB =              -1.603 DEG. F
DIFFERENCE DUE TO DRY BULB TEMP. =         -.371 DEG. F
DIFFERENCE DUE TO WIND SPEED =             2.988 DEG. F
DIFFERENCE DUE TO INSOLATION =             0.000 DEG. F
DIFFERENCE DUE TO BAROMETRIC PRESSURE =    0.000 DEG. F
SUMMATION OF INDIVIDUAL DIFFERENCES =      1.014 DEG. F
```

```
******************************************************************************************

          WET BULB      DRY BULB   WIND SPEED                           POND TEMP    EVAPORATION
          SOLAR RAD.    (DEG,F)      (MPH)    (BTU/FT**2/DY)  PB         (DEG. F)       FT**3
          (DEG. F)                                          INCHES HG

DATA SET 1   62.61       70.60       7.59       1686.60      29.60        80.78      2269915.48

DATA SET 2   55.90       63.10       4.80       1686.60      29.60        78.99      2063438.26

                                               E2-E1 = -1.787      EVAP2-EVAP1 =  -206477.2
```

DIFFERENCES IN E BETWEEN DATA SET 2 AND DATA SET 1 BY PARAMETER

```
DIFFERENCE DUE TO WET BULB =              -3.164 DEG. F
DIFFERENCE DUE TO DRY BULB TEMP. =          .004 DEG. F
DIFFERENCE DUE TO WIND SPEED =             1.045 DEG. F
DIFFERENCE DUE TO INSOLATION =             0.000 DEG. F
DIFFERENCE DUE TO BAROMETRIC PRESSURE =    0.000 DEG. F
SUMMATION OF INDIVIDUAL DIFFERENCES =     -2.115 DEG. F
```

Figure 8.22 (Continued)

****

	WET BULB SOLAR RAD. (DEG. F)	DRY BULB (DEG. F)	WIND SPEED (MPH)	(BTU/FT**2/DY)	PB INCHES HG	POND TEMP (DEG. F)	EVAPORATION FT**3
DATA SET 1	66.46	72.27	7.54	1763.90	29.64	83.02	2294305.86
DATA SET 2	59.10	66.40	4.12	1763.90	29.64	80.52	2231119.09

E2-E1 = -2.497   EVAP2-EVAP1 = -63186.8

DIFFERENCES IN E BETWEEN DATA SET 2 AND DATA SET 1 BY PARAMETER

DIFFERENCE DUE TO WET BULB = -3.724 DEG. F
DIFFERENCE DUE TO DRY BULB TEMP. = -.135 DEG. F
DIFFERENCE DUE TO WIND SPEED = 1.234 DEG. F
DIFFERENCE DUE TO INSOLATION = 0.000 DEG F
DIFFERENCE DUE TO BAROMETRIC PRESSURE = 0.000 DEG F
SUMMATION OF INDIVIDUAL DIFFERENCES = -2.625 DEG. F

****

	WET BULB SOLAR RAD. (DEG. F)	DRY BULB (DEG. F)	WIND SPEED (MPH)	(BTU/FT**2/DY)	PB INCHES HG	POND TEMP (DEG. F)	EVAPORATION FT**3
DATA SET 1	68.34	76.47	5.74	1377.30	29.71	84.47	2353480.48
DATA SET 2	59.50	68.00	3.39	1377.30	29.71	80.39	2148689.17

E2-E1 = -4.080   EVAP2-EVAP1 = -204791.3

DIFFERENCES IN E BETWEEN DATA SET 2 AND DATA SET 1 BY PARAMETER

DIFFERENCE DUE TO WET BULB = -4.428 DEG. F
DIFFERENCE DUE TO DRY BULB TEMP. = -.521 DEG. F
DIFFERENCE DUE TO WIND SPEED = .865 DEG. F
DIFFERENCE DUE TO INSOLATION = 0.000 DEG. F
DIFFERENCE DUE TO BAROMETRIC PRESSURE = 0.000 DEG F
SUMMATION OF INDIVIDUAL DIFFERENCES = -4.084 DEG. F

Figure 8.22 (Continued)

```
****************************************************************************************************

            WET BULB    DRY BULB    WIND SPEED                  PB          POND TEMP    EVAPORATION
            SOLAR RAD.                                                                   FT**3
            (DEG. F)    (DEG.F)     (MPH)      (BTU/FT**2/DY)   INCHES HG    (DEG. F)

DATA SET 1  59.29       64.24       7.46       1182.60          29.70        78.29        2022014.79

DATA SET 2  -53.60      58.50       4.18       1182.60          29.70        78.39        1829890.64

                                                          E2-E1 =  .096    EVAP2-EVAP1 =  -192124.1

****************************************************************************************************

DIFFERENCES IN E BETWEEN DATA SET 2 AND DATA SET 1 BY PARAMETER

    DIFFERENCE DUE TO WET BULB =                   -2.958 DEG. F
    DIFFERENCE DUE TO DRY BULB TEMP. =              -.311 DEG. F
    DIFFERENCE DUE TO WIND SPEED =                  1.508 DEG. F
    DIFFERENCE DUE TO INSOLATION =                  0.000 DEG. F
    DIFFERENCE DUE TO BAROMETRIC PRESSURE =         0.000 DEG. F
    SUMMATION OF INDIVIDUAL DIFFERENCES =          -1.761 DEG. F

****************************************************************************************************

            WET BULB    DRY BULB    WIND SPEED                  PB          POND TEMP    EVAPORATION
            SOLAR RAD.                                                                   FT**3
            (DEG. F)    (DEG.F)     (MPH)      (BTU/FT**2/DY)   INCHES HG    (DEG. F)

DATA SET 1  59.57       64.74       6.65       1559.60          29.61        79.49        2052606.74

DATA SET 2  54.60       61.00       4.36       1559.60          29.61        75.77        1977817.37

                                                          E2-E1 = -.723    EVAP2-EVAP1 =   -74789.4

DIFFERENCES IN E BETWEEN DATA SET 2 AND DATA SET 1 BY PARAMETER

    DIFFERENCE DUE TO WET BULB =                   -2.197 DEG. F
    DIFFERENCE DUE TO DRY BULB TEMP. =               .396 DEG. F
    DIFFERENCE DUE TO WIND SPEED =                   .929 DEG. F
    DIFFERENCE DUE TO INSOLATION =                  0.000 DEG. F
    DIFFERENCE DUE TO BAROMETRIC PRESSURE =         0.000 DEG. F
    SUMMATION OF INDIVIDUAL DIFFERENCES =           -.872 DEG. F
```

Figure 8.22 (Continued)

```
*************************************************************************************************

            WET BULB
            SOLAR RAD.   DRY BULB   WIND SPEED                   PB         POND TEMP   EVAPORATION
            (DEG. F)     (DEG.F)    (MPH)      (BTU/FT**2/DY)   INCHES HG   (DEG. F)    FT**3

DATA SET 1    65.36      70.57       7.61        1636.90         29.67       82.13      2223929.43

DATA SET 2    58.00      65.70       4.53        1636.90         29.67       79.71      2125769.65


DIFFERENCES IN E BETWEEN DATA SET 2 AND DATA SET 1 BY PARAMETER

   DIFFERENCE DUE TO WET BULB =                 -3.664 DEG. F
   DIFFERENCE DUE TO DRY BULB TEMP. =            -.008 DEG. F
   DIFFERENCE DUE TO WIND SPEED =                1.103 DEG. F
   DIFFERENCE DUE TO INSOLATION =                0.000 DEG. F
   DIFFERENCE DUE TO BAROMETRIC PRESSURE =       0.000 DEG. F
   SUMMATION OF INDIVIDUAL DIFFERENCES =        -2.569 DEG. F

*************************************************************************************************

            WET BULB
            SOLAR RAD.   DRY BULB   WIND SPEED                   PB         POND TEMP   EVAPORATION
            (DEG. F)     (DEG.F)    (MPH)      (BTU/FT**2/DY)   INCHES HG   (DEG. F)    FT**3

DATA SET 1    69.36      75.01       6.84        1746.90         29.65       85.05      2357392.80

DATA SET 2    60.30      69.40       3.54        1746.90         29.65       81.30      2244560.67

                                                          E2-E1 = -3.743    EVAP2-EVAP1 =  -112832.1

DIFFERENCES IN E BETWEEN DATA SET 2 AND DATA SET 1 BY PARAMETER

   DIFFERENCE DUE TO WET BULB =                 -4.692 DEG. F
   DIFFERENCE DUE TO DRY BULB TEMP. =            -.329 DEG. F
   DIFFERENCE DUE TO WIND SPEED =                1.183 DEG. F
   DIFFERENCE DUE TO INSOLATION =                0.000 DEG. F
   DIFFERENCE DUE TO BAROMETRIC PRESSURE =       0.000 DEG. F
   SUMMATION OF INDIVIDUAL DIFFERENCES =        -3.839 DEG. F

*************************************** Figure 8.22  (Continued) ************************************
```

8-59

	WET BULB SOLAR RAD. (DEG. F)	DRY BULB (DEG,F)	WIND SPEED (MPH)	(BTU/FT**2/DY)	PB INCHES HG	POND TEMP (DEG. F)	EVAPORATION FT**3
DATA SET 1	69.63	75.12	6.75	1507.40	29.70	84.90	2322665.82
DATA SET 2	59.60	68.20	3.94	1507.40	29.70	80.41	2176958.91
					E2-E1 = -4.485	EVAP2-EVAP1 =	-145706.9

DIFFERENCES IN E BETWEEN DATA SET 2 AND DATA SET 1 BY PARAMETER

DIFFERENCE DUE TO WET BULB = -5.182 DEG. F
DIFFERENCE DUE TO DRY BULB TEMP. = -.354 DEG. F
DIFFERENCE DUE TO WIND SPEED = -.976 DEG. F
DIFFERENCE DUE TO INSOLATION = 0.000 DEG. F
DIFFERENCE DUE TO BAROMETRIC PRESSURE = 0.000 DEG. F
SUMMATION OF INDIVIDUAL DIFFERENCES = -4.560 DEG. F

****

	WET BULB SOLAR RAD. (DEG. F)	DRY BULB (DEG,F)	WIND SPEED (MPH)	(BTU/FT**2/DY)	PB INCHES HG	POND TEMP (DEG. F)	EVAPORATION FT**3
DATA SET 1	58.99	62.82	7.31	1157.00	29.76	78.22	1973356.18
DATA SET 2	53.20	57.90	4.47	1157.00	29.76	78.21	1807507.15
					E2-E1 = -.005	EVAP2-EVAP1 =	-165849.0

DIFFERENCES IN E BETWEEN DATA SET 2 AND DATA SET 1 BY PARAMETER

DIFFERENCE DUE TO WET BULB = -2.978 DEG. F
DIFFERENCE DUE TO DRY BULB TEMP. = -.263 DEG. F
DIFFERENCE DUE TO WIND SPEED = 1.418 DEG. F
DIFFERENCE DUE TO INSOLATION = 0.000 DEG. F
DIFFERENCE DUE TO BAROMETRIC PRESSURE = 0.000 DEG. F
SUMMATION OF INDIVIDUAL DIFFERENCES = -1.823 DEG. F

****

SAMPLE R SQUARED FOR EQUILIBRIUM TEMP. = .755   .843   STANDARD ERROR = .506 DEG.F
SAMPLE R SQUARED FOR EVAPORATION = .755   .843   STANDARD ERROR = 83405.079FT**3
AVERAGE E, DATA SET 1 = 81.061
AVERAGE E, DATA SET 2 = 79.701
AVERAGE E2 - AVERAGE E1 = -1.3596
AVERAGE EVAP2 - AVERAGE EVAP1 = -163463.6920

Figure 8.22 (Continued).

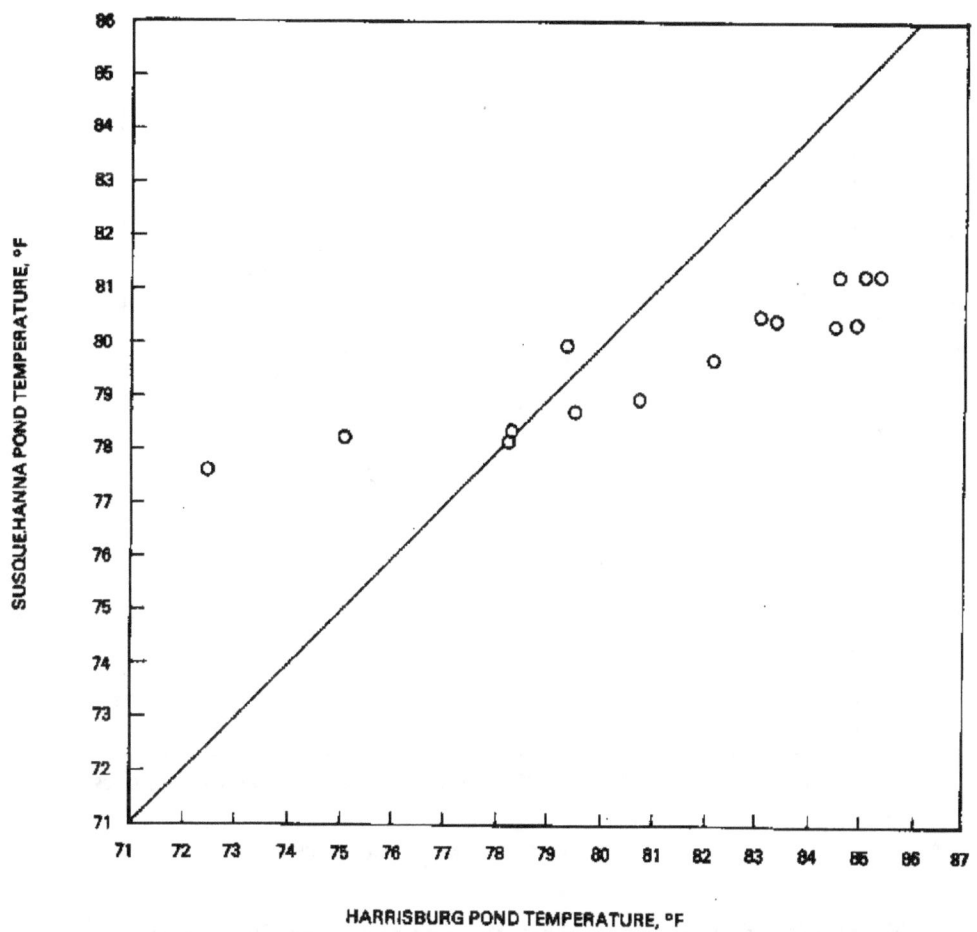

Figure 8.23   Comparison of Susquehanna site and Harrisburg spray-pond
temperatures, program COMET2

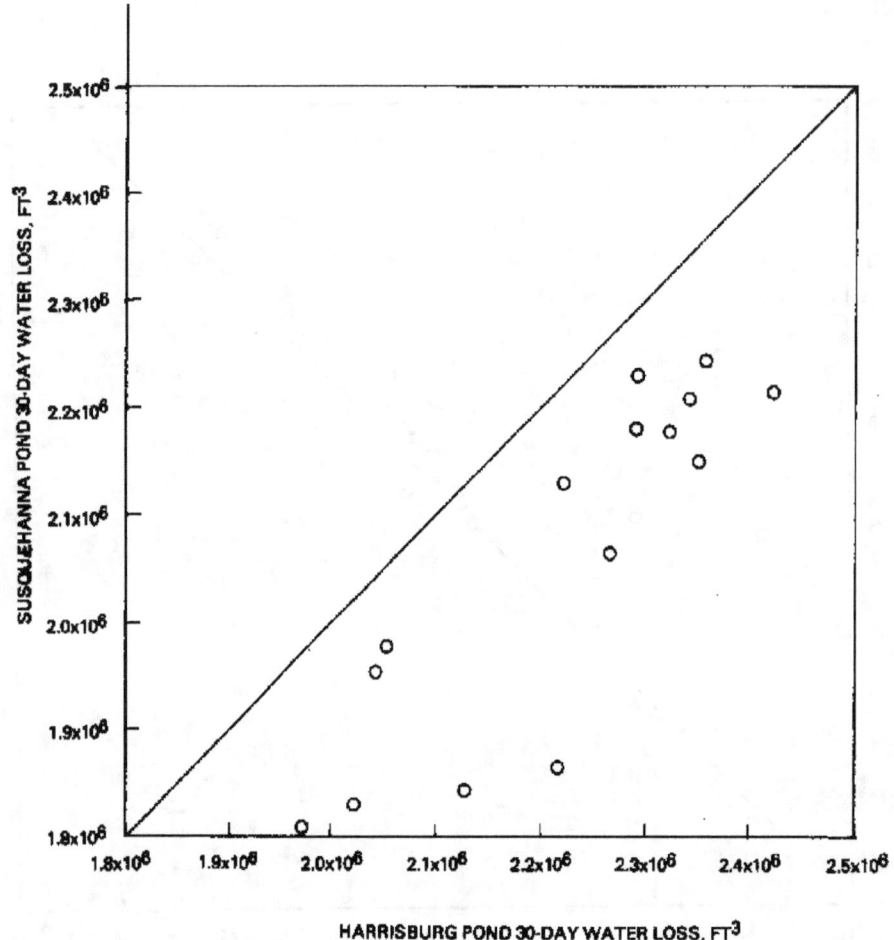

Figure 8.24   Comparison of Susquehanna site and Harrisburg pond
water losses, program COMET2

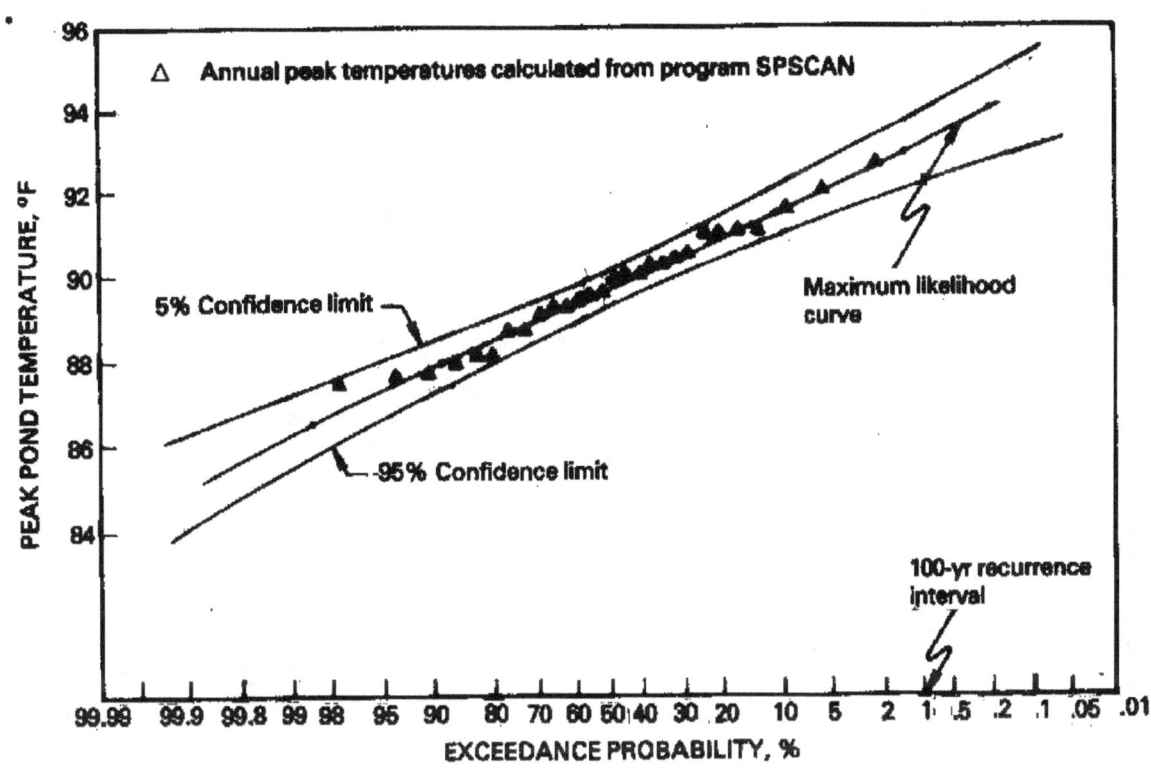

Figure 8.25 Exceedance probability for annual peak pond temperature

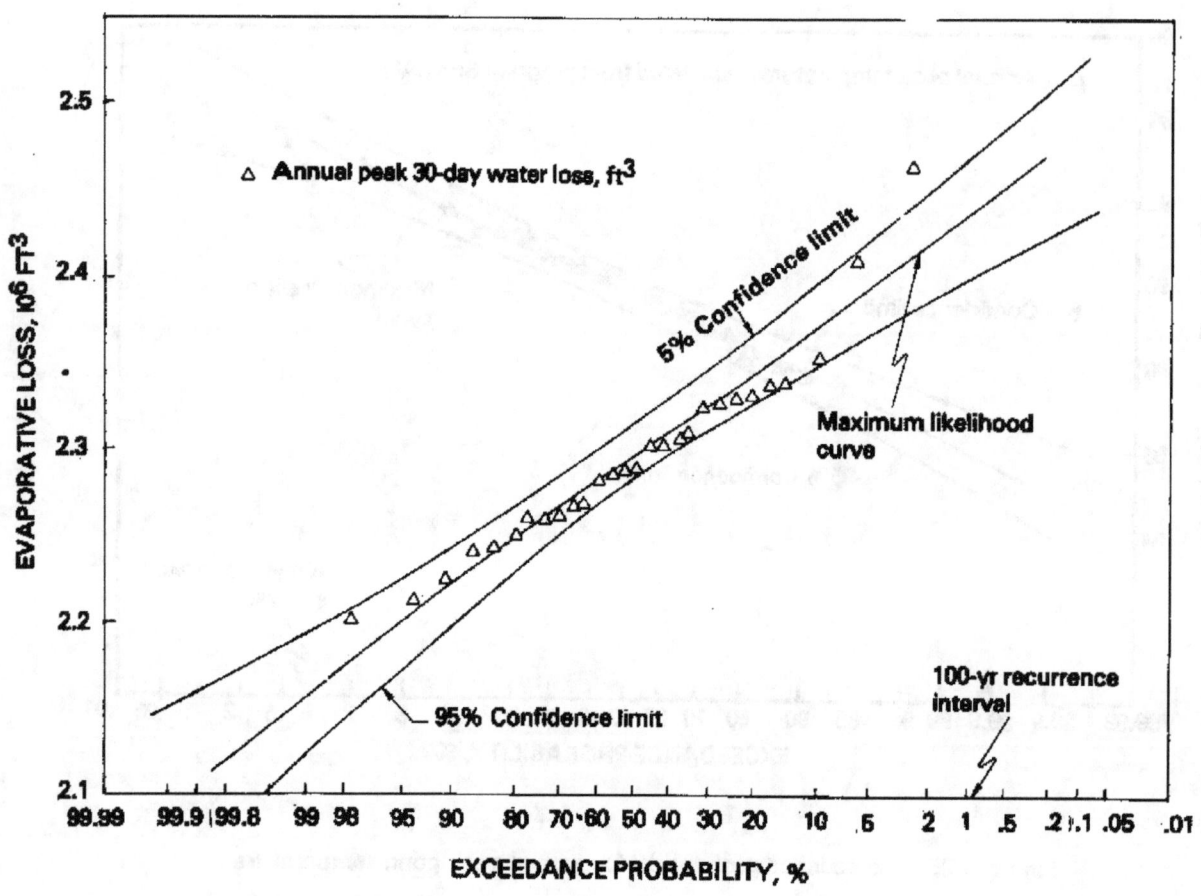

Figure 8.26 Exceedance probability for annual peak 30-day.pond water loss

9 REFERENCES

1. U.S. Nuclear Regulatory Commission, Regulatory Guide 1.27, "Ultimate Heat Sink for Nuclear Power Plants."*

2. A. M. Elgawhary, "Spray Pond Mathematical Model for Cooling Fresh Water and Brine," Ph.D. Thesis, Oklahoma State University, Stillwater, 1971.

3. Pennsylvania Power and Light, "Preliminary Safety Analysis Report, Susquehanna Steam Electric Station Units 1 and 2," Docket 50-387.**

4. R. B. Bird, W. E. Stewart, and E. N. Lightfoot, Transport Phenomena, John Wiley and Sons, Inc., New York, NY, 1960.

5. W. D. Ranz and W. R. Marshall, "Evaporation From Drops," Chemical Engineering Progress 48, 141-173 (1952).***

6. D. M. Himmelblau, Basic Principles and Calculations in Chemical Engineering, Prentice Hall, Inc., Englewood Cliffs, NJ, 1962.

7. Ecolaire Co., "Topical Report-Oriented Spray Cooling System," Ecolaire Condenser, Lehigh Valley, PA, January 1977.

8. R. B. Codell, "Digital Computer Simulation of Thermal Effluent Dispersion in Rivers, Lakes and Estuaries," Ph.D. Thesis, Lehigh University, Bethlehem, PA, 1973.

9. G. W. Branne, "Spray Pond Testing at Canadys Station," South Carolina Electric and Gas Co., Columbia, SC, unpublished.**

See footnotes at end of list.

10. V. E. Schrock and G. J. Trezek, "Rancho Seco Nuclear Service Spray Ponds Performance Evaluation," Report No. WHM-4, University of California (Berkeley) for Sacramento Municipal Utility District, July 1973.

11. SPRACO, "Spray Ponds--The Answer to Thermal Pollution Problems," Spray Engineering Company, Burlington, MA, undated.

12. K. H. Chen and G. J. Trezek, "Spray Energy Release (SER) Approach to Analyzing Spray System Performance," Proceedings of American Power Conference, Vol. 38, 1435-1448, Illinois Institute of Technology, Chicago, IL, 1976.

13. J. L. Baldwin, Climates of the United States, U.S. Department of Commerce, National Oceanic and Atmospheric Administration, Washington, DC, 1973.

14. R. B. Codell and W. K. Nuttle, "Analysis of Ultimate Heat Sink Cooling Ponds," USNRC Report NUREG-0693, November 1980.†

15. G. H. Jirka, G. Abraham, and D. R. F. Harleman, "An Assessment of Techniques of Hydrothermal Prediction," Technical Report No. 203, R. M. Parsons Laboratory for Water Resources and Hydrodynamics, Department of Civil Engineering, Massachusetts Institute of Technology, Cambridge, MA, July 1975 (also available as NUREG-0044†, March 1976).

16. P. J. Ryan and D. R. F. Harleman, "An Analytical and Experimental Study of Transient Cooling Ponds Behavior," Report No. 161, R. M. Parsons Laboratory for Water Resources and Hydrodynamics, Department of Civil Engineering, Massachusetts Institute of Technology, Cambridge, MA, January 1973.

17. D. K. Brady, W. L. Graves, and J. C. Geyer, Surface Heat Exchange at Power Plant Cooling Lakes, Report No. 5, Edison Electric Institute, EEI Publication 69-401, New York, NY, 1969.

See footnotes at end of list.

18. R. W. Hamon, L. L. Wiess, and W. T. Wilson, "Insolation as an Empirical Function of Daily Sunshine Duration," Monthly Weather Review 82(6):141-146, June 1954.***

19. W. O. Wunderlich, "Heat and Mass Transfer Between a Water Surface and the Atmosphere," Report No. 14, TVA Engineering Laboratory, Norris, TN, 1972.

20. L. R. Beard, "Statistical Methods in Hydrology," Engineering Memorandum EM 1110-2-1450, U.S. Army Engineer District, Corps of Engineers, Sacramento, CA, January 1962.

---

*Single copies of Active Guides are available for $1.50 each. Send check/ money order (made payable to Superintendent of Documents) to the U.S. Nuclear Regulatory Commission, Washington, DC 20555. ATTN: Sales Manager.

**Available in NRC Public Document Room (1717 H St., N.W., Washington, DC 20555) for inspection and copying for a fee.

***Available in public technical libraries.

†Available for purchase from the NRC/GPO Sales Program, U.S. Nuclear Regulatory Commission, Washington, DC 20555, and/or the National Technical Information Service, Springfield, VA 22161.

APPENDIX A

STATISTICAL TREATMENT OF OUTPUT

Program SPSCAN, in addition to determining the peak ambient pond temperature for the entire length of record, determines the maximum ambient temperature and evaporation for each year of the record and performs several manipulations of the yearly maximums to facilitate graphic analyses:

(1) The data are ranked from highest to lowest temperature.

(2) Their "probability" or plotting position is determined based on the number of years in the data set using the formulae (Ref. 20):

$$P_1 = 1 - (0.5)^{1/N} \tag{A.1}$$

$$P_N = (0.5)^{1/N} \tag{A.2}$$

$$P_i = P_1 - (i-1)\Delta P \tag{A.3}$$

where

$$\Delta P = \frac{2\left(0.5 - P_1\right)}{N-1}$$

N = number of data points in the set

P_1 = plotting position of the highest yearly maximum

P_N = plotting position of the lowest yearly maximum

P_i = plotting position of each individual point

(3) The first two moments of the distribution of any variable T (mean and standard deviation) are determined from the formulae (Ref. 20):

$$M = \frac{\Sigma T}{N} \qquad\qquad \text{(sample mean)} \qquad\qquad \text{(A.4)}$$

$$s^2 = \frac{\Sigma T^2 - (\Sigma T)^2/N}{N - 1} \qquad\qquad \text{(standard deviation)}^2 \qquad \text{(A.5)}$$

where

Σ implies the sum over all N values in the data set

(4) The maximum likelihood curve and confidence limits of temperature and water loss are calculated. The probabilities of the data are assumed to be representable by Student's t distribution.

A.1 Maximum Likelihood Curve

The maximum likelihood frequency curve for any variable T in probability coordinates is described by the equation:

$$T = M + sk \qquad\qquad\qquad\qquad \text{(A.6)}$$

where

M = sample mean of T

s = standard deviation of T

and

k = the $100 (1 - P)^{th}$ percentile of Student's t distribution with N - 1 degrees of freedom,

where P = the probability (independent variable)

N = the sample size.

A.2 Confidence Limits

The 5% and 95% confidence limits of T are calculated from the formulae

$$T_{95} = T + \sqrt{\frac{s^2}{N}\left(1 + \frac{k^2}{2}\right)} \qquad k \qquad\qquad \text{(A.7)}$$

$$T_5 = T - \sqrt{\frac{s^2}{N}\left(1 + \frac{k^2}{2}\right)} \quad k \tag{A.8}$$

The 95% and 5% confidence limits and maximum likelihood curve are calculated for probabilities P ranging from 0.001 to 0.999. These points should be plotted as smooth curves on probability-scale paper along with the ranked raw data.

The error-limit curves express the probability of a value falling outside of the error banks in any given year. For the 95% and 5% bands, therefore, there is 1 chance in 20 that the ambient temperature value for any given recurrence interval is greater than indicated by the 5% curve and 1 chance in 20 that it is less than the 95% curve.

The conservatism of choosing the design-basis event coincident with the most adverse meteorological conditions may be demonstrated with the following procedure.

The maximum-likelihood curves for temperature T (°F) and 30-day evaporation, may be extrapolated to the 100-yr recurrence intervals (0.01 probability per year) T_{100} and W_{100}, respectively, or to any other justifiable recurrence interval. Correction factors for peak pond temperature ΔT and evaporation ΔW_e are determined by comparing T_{100} and W_{100} with their corresponding highest observed values from the record, T_{max} and W_{max}:

$$\Delta T = T_{100} - T_{max} \text{ °F} \tag{A.10}$$

$$\Delta W_e = W_{100} - W_{max} \text{ ft}^3/30 \text{ days} \tag{A.11}$$

Only correction factors greater than zero should be considered. If the maximum observed temperature or evaporation is higher than the 100-yr (or other period) recurrence values, no correction factor is taken. These correction factors may be added directly to the peak loaded pond temperature and evaporations determined in subsequent calculations.

An example of the statistical procedure is offered in Section 7.

APPENDIX B

COMPUTER CODES

```
      PROGRAM SPRCO(INPUT,OUTPUT,TAPE7,TAPE9,TAPE5=INPUT,TAPE6=OUTPUT,
     1 PUNCH,TAPE4=PUNCH)
C     SPRAY POND CORRELATION MODEL
C     RICHARD CODELL
C     U.S. NUCLEAR REGULATORY COMMISSION, WASHINGTON D.C.
C     GENERATES A SET OF PERFORMANCES FROM THE HIGH WIND SPEED(HWS) MODEL
C     AND THE LOW WIND SPEED MODEL(LWS) AND CORRELATES THE RESULTS TO
C     A SET OF MULTILE LINEAR REGRESSION EQUATIONS
      DIMENSION TSEG(11),HUM(11)                                          000130
      COMMON/LWSCOM/ ATOP(12),ASIDE(12)                                  000140
      COMMON A,VOL,AM,CON1,CON2,CON3,CON4,CON5,VIS,RHOA,DIFF,            000160
     1 PR,AK,DT,H,EVAP,NSTEPS,CON6,DTO6,DTO2,TDROP                       000170
     1 ,UO,VO,SC                                                        000180
      C(Z)=(Z-32)/1.8                                                    000190
C     ALPHA IS CONVERGENCE PARAMETER OF LWS MODEL
      DATA ALPHA/=.2/
      DATA NPNTS,VELO,THETA,YO,R,PB,PHI/200.22.5,71.0,5.0,.104,
     1 29.92,90.0/
      DATA TWETO,DTDRYO,WINDO,THOTO,RTW,RTD,RW,RTH/50.0,20.0,
     1 0.1,90.0,30.0,30.0,20.0,30.0/
      NAMELIST/INPUT/ NPNTS,HT,ALEN,WID,VELO,THETA,YO,R,PB,
     1 Q,PHI,TWETO,DTDRYO,WINDO,THOTO,RTW,RTD,RW,RTH
      REWIND 7
      REWIND 9                                                          000230
      WRITE(6,22)
   22 FORMAT(1H1,30X,'COEFFICIENTS FOR EFFICIENCY AND EVAPORATION'/
     1 10X,'FROM A SPRAY FIELD'///30X,'INPUT VARIABLES')
      READ(5,INPUT)
      WRITE(6,200) NPNTS,VELO,THETA,R,PB,HT,WID,ALEN,YO,Q,PHI
  200 FORMAT(///,20X,'NUMBER OF RANDOM POINTS,NPNTS = ',I5/
     1 20X,'INITIAL VELOCITY OF DROPS LEAVING NOZZLE, VELO = ',F10.2,
     2 ' FT/SEC'/
     3 20X,'INITIAL ANFLE OF DROPS TO HOR., THETA = ',F10.3,' DEGREES'/
     4 20X,'GEOMETRIC MEAN RADIUS OF DROPS, R = ',F10.4,' CM'/
     5 20X,'ATMOSPHERIC PRESSURE, PB = ',F10.2,' INCHES HG'/
     6 20X,'HEIGHT OF SPRAY FIELD, HT = ',F10.2,' FT'/
     7 20X,'WIDTH OF SPRAY FIELD, WID = ',F10.1,' FT'/
     8 20X,'LENGTH OF SPRAY FIELD, ALEN = ',F10.1,' FT'/
     8 20X,'HEIGHT OF SPRAY NOZZLES ABOVE POND SURFACE, YO = ',F10.1,
     * ' FT'/20X,'FLOWRATE OF WATER SPRAYED, Q = ',F10.2,' CU.FT./SEC',/
     * 20X,'HEADING OF WIND W.R.T.LONG AXIS, PHI = ',F10.2,' DEGREES'//)
C     CONVERT SPRAY PARAMETERS TO METRIC UNITS
      VELO=VELO*30.48
      THETA =THETA*(3.1415926/180.0)
      HT=HT*30.48
      WID=WID*30.48
      ALEN=ALEN*30.48
      YO=YO*30.48
      Q=Q*28316
      TWETH=TWETO+RTW
      TDRYO=TWETO+DTDRYO
      TDRYH=TWETH+RTD+DTDRYO
      WH=WINDO+RW
      THOTH=THOTO+RTH
      WRITE(6,201) TWETO,TWETH,TDRYO,TDRYH,WINDO,WH,THOTO,THOTH
  201 FORMAT(/40X,'RANGES OF METEOROLOGICAL PARAMETERS'/20X,
     1 'WET BULB TEMPERATURE = ',F10.3,' TO',F10.3,' DEG.F'/20X,
     3 'DRY BULB TEMPERATURE = ',F10.3,' TO',F10.3,' DEG.F'/ 20X,
     2 'WIND SPEED = ',F10.3,' TO',F10.3,' MPH'/20X,
     3 'SPRAYED TEMPERATURE = ',F10.3,' TO',F10.3,' DEG.F'//)
C     NPNTS = THE NUMBER OF POINTS IN THE CORRELATION
C     HT = THE HEIGHT OF THE SPRAY FIELD, FT
```

Figure B.1   Listing of program SPRCO

B-2

```
C     ALEN = THE LENGTH OF THE SPRAY FIELD, FT
C     WID = THE WIDTH OF THE SPRAY FIELD, FT
C     VELO = THE INITIAL VELOCITY OF THE DROPS LEAVING THE NOZZLE, FT/SEC
C     THETA = THE ANGLE OF THE DROPS LEAVING THE NOZZLE W.R.T. HORIZON, DEGREES
C     YO = THE HEIGHT OF THE NOZZLES ABOVE POND SURFACE, FT
C     R = THE GEOMETRIC MEAN DROP SIZE, CM
C     PB = BAROMETRIC PRESSURE,INCHES MERCURY
C     Q = QUANTITY OF WATER SPRAYED THROUGH FIELD, CUBIC FEET PER SECOND
C     TWETO=LOWER LIMIT OF RANGE OF WET BULB T.,F
C     DTDRYO = LOWER LIMIT ON RANGE OF DRY BULB T ADDED TO WET BULB T, F
C     WINDO = LOWER LIMIT OF WIND SPEED RANGE,MPH
C     THOTO = LOWER LIMIT OF SPRAYED WATER TEMPERATURE, F
C     RTW = RANGE OF WET BULB TEMPERATURE, F
C     RTD = RANGE OF DRY BULB TEMPERATURE, F
C     RW = RANGE OF WIND SPEED, MPH
C     RTH = RANGE OF SPRAYED TEMPERATURE, F
      WRITE(9) NPNTS
C     AREA OF SIDE OF SPRAY POND IN HWS MODEL
      ASIDEH=HT*ALEN                                                        000300
      NSTEPS=10
      DLEN=ALEN/10                                                          000340
      DWID=WID/10                                                           000350
      DO 801 J=1,10                                                         000360
      I=12-J                                                                000370
C     TOP AND SIDE AREAS FOR EACH SEGMENT IN LWS MODEL
      ATOP(I)=J*DLEN*DWID-J*(J-1)*DLEN*DWID*(J-1)                           000380
      ASIDE(I)=((J-1)*DLEN+(J-1)*DWID)*2*HT                                 000390
  801 CONTINUE                                                             000400
      ASIDE(1)=(ALEN+WID)*2*HT                                            000410
      ASIDE(12)=0                                                         000420
      CALL INIT(R,THETA,YO,VELO)                                          000430
      WRITE(6,6)
    6 FORMAT(10X,'PT NO.',T20,'TWET',T30,'TDRY',T40,'THOT',T50,'WIND',
     1 T61,'HUMID',T71,'ETA',T81,'ETA',T92,'EVAP.',T105,'EVAP.',
     2 /T22,'F',T32,'F',T42,'F',T51,'MPH',T92,'LWS',T105,'HWS',
     3 T71,'LWS',T81,'HWS'/)
      DO 1 I=1,NPNTS                                                        000520
C     GENERATE RANDOM MET DATA
      CALL RANDIN(TWETO,DTDRYO,WINDO,THOTO,RTW,RTD,RW,                     000530
     1 RTH,TWET,TDRY,WIND,THOT,PB)                                         000540
      WIND=WIND*ABS(SIN(PHI*.017453293))                                  000560
C  CONVERT MPH TO CM/SEC                                                   000570
      WIND1=WIND*44.7
C     CALCULATE HUMIDITY
      CALL PSY1(TDRY,TWET,PB,DP,PV,HUMID,ENTHAL,VOLUME,RH)                000590
      THOT1=C(THOT)
      TDRY1=C(TDRY)
      TWET1=C(TWET)
C      HIGH WIND SPEED MODEL
C    , USE HIGH WIND SPEED MODEL
      CALL HWS(THOT1,HUMID,TDRY1,ASIDEH,TWAV,WIND1,Q,R,EVAPS)
C     HWS EFFICIENCY AND EVAPORATION
      ETA2=(THOT1-TWAV)/(THOT1-TWET1)
      ETA2S=ETA2                                                           000670
      EVAPSS=EVAPS/Q                                                       000680
C     DELIBERATELY SET TO EXCEED FORMAT, THEREBY PRINTING STARS
      ETA2=-9999                                                           000700
      EVAPS=-9999999
      IF(TDRY.GT.THOT) GOTO 1111                                          000710
      DO 444 L=2,11                                                        000720
      TSEG(L)=TDRY1+1.0
```

Figure B.1   (Continued)

```
  444 HUM(L)=HUMID+.01                                              000740
    5 FORMAT(10X,I5,5F10.4,3X,F7.4,F10.4,2(5X,F9.6))
C     LOW WIND SPEED MODEL
      CALL LWS(THOT1,HUMID,TDRY1,ALEN,WID,TWAV,Q,R,
    1 TSEG,HUM,ALPHA,HT,EVAPS)                                       000770
C     LWS EFFICIENCY AND EVAPORATION
      ETA2=(THOT1-TWAV)/(THOT1-TWET1)
      EVAPS=EVAPS/Q                                                  000810
 1111 CONTINUE                                                       000820
      WRITE(9) TWET,TDRY,THOT,WIND,HUMID,ETA2,
    1 ETA2S,EVAPS,EVAPSS
      WRITE(6,5) I,TWET,TDRY,THOT,WIND,HUMID,ETA2,ETA2S,EVAPS,EVAPSS
    1 CONTINUE                                                       000860
C     GENERATE REGRESSION EQUATIONS
      CALL FITSPR                                                    000865
      STOP                                                           000870
      END                                                           000880
      SUBROUTINE RANDIN(TWET0,DTDRY0,WIND0,THOT0,RTW,RTD,            000890
    1 RW,RTH,TWET,TDRY,WIND,THOT,PB)                                 000900
C     GENERATES RANDOM VALUES OF METEOROLOGICAL VARIABLES
      DO 1 I=1,10                                                    000910
      TWET=TWET0+RTW*RANF(J)                                         000920
      TDRY=TWET+DTDRY0+RTD*RANF(J)                                   000930
C     CHECK FOR PLAUSIBILITY OF TWET WITH RESPECT TO TDRY
      CALL PSY1(TDRY,TWET,PB,DP,PV,HUMID,H,V,RH)                     000940
      IF(HUMID.GT.0) GOTO 2                                          000950
      GOTO 1
    2 WIND=WIND0+RW*RANF(J)                                          000970
      THOT=THOT0+RTH*RANF(J)                                         000980
      IF(THOT.LE.TDRY.AND.WIND.LT.1.0) GOTO 1
      GOTO 3
    1 CONTINUE                                                       000960
    3 CONTINUE
      RETURN                                                         000990
      END                                                           001000
      SUBROUTINE LWS(THOT,HUMID,TAIR,ALEN,WID,TWAV,Q,R,TSEG,        001010
    1 HUM,ALPHA,HT,EVAPS)                                            001020
C     LOW WIND SPEED MODEL
      COMMON A,VOL,AM,CON1,CON2,CON3,CON4,CON5,VIS,                  001030
    1 RHOA,DIFF,PR,AK,DT,H,EVAP,NSTEPS,CON6,DTO6,                    001040
    2 DTO2,TDROP,U0,V0,SC                                            001050
      COMMON/LWSCOM/ ATOP(12),ASIDE(12)                             001060
      DIMENSION VUP(12),FLOW(12),QT(12),RHO2(12),VH(12)
      DIMENSION TSEG(11),HUM(11),HOUT(11)                            001090
      DIMENSION HFIL(12),TFIL(12)                                    001100
      DIMENSION TM2(12),TM1(12),HM2(12),HM1(12)                     001110
      DO 491 I=1,12                                                  001120
      TM2(I)=0                                                       001130
      TM1(I)=0                                                       001140
      HM2(I)=0                                                       001150
  491 HM1(I)=0                                                       001160
      TLAST=0                                                        001170
      DATA HVAP,CP,RHO/580.0,1.0,1.0/                                001180
      ICNT=0                                                         001190
C     DENSITY OF AMBIENT AIR GM/CC
      RHO1=(1+HUMID)/((81.86*TAIR+22387)*(.03448+HUMID/18))          001200
      FLOW(11)=0                                                     001210
      QT(1)=0                                                        001220
      FLOW(1)=0                                                      001230
      RHO2(1)=RHO1                                                   001240
      ATOT=ALEN*WID                                                  001250
      TSEG(1)=TAIR                                                   001260
```

Figure B.1   (Continued)

```
      HUM(1)=HUMID                                                      00127C
C     CONCENTRATION OF WATER IN AIR
      CWA=HUMID/((81.86*TAIR+22387)*(.03448+HUMID/18))                  001280
C     BEGIN ITERATIVE SOLUTION
      DO 801 NITER=1,20                                                 001290
      DO 101 J=1,10                                                     001300
      I=12-J                                                            001310
C     DENSITY OF AIR IN EACH SEGMENT GM/CC
      RHO2(I)=(1+HUM(I))/((81.86*TSEG(I)+22387)*(.03448+HUM(I)/18))     001320
C     HUMID VOLUME, CC/GM BDA
      VH(I)=((81.86*TSEG(I)+22387)*(.03448+HUM(I)/18))                  001330
  101 CONTINUE                                                          001340
  105 CONTINUE                                                          001350
      DO 1001 J=1,10                                                    001360
      I=12-J                                                            001370
      DRHO=RHO1-RHO2(I)                                                 001380
      ARG=980*DRHO*HT*.5/RHO1                                           001390
      ICNT=1                                                            001400
      IF(ARG.LT.0.0) GOTO 668                                          001410
C     UPWARD VELOCITY OF AIR LEAVING EACH SEGMENT
      VUP(I)=SQRT(ARG)                                                  001420
  668 CONTINUE                                                          001430
C     MATERIAL BALANCE ON EACH SEGMENT
      QT(I)=VUP(I)*ATOP(I)/VH(I)                                        001440
      FLOW(I-1)=FLOW(I)+QT(I)                                           001450
 1001 CONTINUE                                                          001460
      ICNT=ICNT+1                                                       001470
  104 CONTINUE                                                          001480
C     ENTHALPY OF AIR ENTERING FIRST SEGMENT, CAL/GM BDA
      HOUT(1)=FLOW(1)*(.238*TAIR+HUMID*(HVAP+.45*TAIR))                 001490
      TSEG(1)=TAIR                                                      001510
      EVAPS=0                                                           001520
      HUM(1)=HUMID                                                      001530
      SUMTC=0                                                           001540
      DO 201 I=2,11                                                     001550
      TEMP=TSEG(I-1)+273.2                                              001560
C     VISCOSITY OF AIR, GM/(SEC CM)
      VIS=2.7936E-6*TEMP**.73617                                        001570
C     DENSITY OF AIR, GM/CC
      RHOA=.353/TEMP                                                    001580
C     DIFFUSION COEFF OF AIR(CM**2/SEC)
      DIFF=5.8758E-6*TEMP**1.8615
C     PRANTL NO
      PR=.93176*TEMP**(-.042784)                                       001600
C     SCHMIDT NO
      SC=2.2705*TEMP**(-.21398)                                        001610
C     THERMAL CONDUCTIVITY OF AIR,CM/SEC
      AK=3.9273E-7*TEMP**.88315                                         001620
      CON4=AK/R                                                         001630
      CON6=2*R*RHOA/VIS                                                 001640
      CON5=DIFF/R                                                       001650
      TDROP=THOT                                                        001660
C     CALCULATE TEMPERATURE AND EVAPORATION OF FALLING DROPS
      CALL DROP(TSEG(I-1),CWA)                                          001670
C     SENSIBLE HEAT TRANSFER IN SEGMENT
      HSEG=RHO*CP*(Q*ATOP(I)/ATOT)*(THOT-TDROP)                        001680
C     EVAPORATION IN SEGMENT
      EVAP1=EVAP*Q*ATOP(I)/(ATOT*VOL)                                  001690
C     SENSIBLE HE AT LEAVING SEGMENT AND ENTERING NEXT
      HOUT(I)=HSEG+HOUT(I-1)*(1-QT(I-1)/(QT(I-1)+FLOW(I-1)))           001700
C     HUMIDITY IN SEGMENT
      HUM(I)=HUM(I-1)+EVAP1/FLOW(I-1)                                  001710
```

Figure B.1 (Continued)

```
C        TEMPERATURE IN SEGMENT
         TSEG(I)=(HOUT(I)/FLOW(I-1)-HUM(I)*HVAP)/(.238+.45*HUM(I))    001720
         EVAPS=EVAPS+EVAP1                                            001730
         CWA=HUM(I)/((81.86*TSEG(I)+22387)*(.03448+HUM(I)/18))        001740
         SUMTC=SUMTC+TDROP*ATOP(I)                                    001750
  201 CONTINUE                                                        001760
C        AVERAGE TEMPERATURE OF WATER FALLING TO POND SURFACE
      TWAV=SUMTC/ATOT                                                 001770
      IF(NITER.LT.3) GOTO 49                                          001790
      DO 492 I=2,11                                                   001800
C        SECOND ORDER SMOOTHING OPERATOR TO AID CONVERGENCE
      HFIL(I)=ALPHA*(HM2(I)-2*HM1(I)+HUM(I))                          001810
      TFIL(I)=ALPHA*(TM2(I)-2*TM1(I)+TSEG(I))                         001820
  492 CONTINUE                                                        001830
      DO 493 I=2,11                                                   001840
      TSEG(I)=TSEG(I)+TFIL(I)                                         001850
      HUM(I)=HUM(I)+HFIL(I)                                           001860
  493 CONTINUE                                                        001870
   49 DO 494 I=2,11                                                   001880
      TM2(I)=TM1(I)                                                   001890
      TM1(I)=TSEG(I)                                                  001900
      HM2(I)=HM1(I)                                                   001910
  494 HM1(I)=HUM(I)                                                   001920
      IF(ABS((TLAST-TWAV)/TWAV).LT.0.002) GOTO 800                    001930
      TLAST=TWAV                                                      001940
  801 CONTINUE                                                        001950
      WRITE(6,20)
   20 FORMAT(10X,'NO CONVERGENCE AFTER 20 TRIES')
  800 RETURN                                                          001970
      END                                                             001980
      SUBROUTINE HWS(THOT,HUMID,TAIR,ASIDE,TWAV,                      001990
     1 WIND,Q,R,EVAPS)                                                002000
C        HIGH WIND SPEED MODEL
      COMMON A,VOL,AM,CON1,CON2,CON3,CON4,CON5,VIS,RHOA,DIFF,         002010
     1 PR,AK,DT,H,EVAP,NSTEPS,CON6,DTO6,DTO2,TDROP                    002020
     1 ,UO,VO,SC                                                      002030
      DIMENSION TSEG(11),HUM(11),HOUT(11)                             002040
      DATA HVAP,CP,RHO/580.0,1.0,1.0/                                 002050
      CON7=RHO*CP*Q/10                                                002060
      CON8=Q/(10*VOL)                                                 002070
C        GMS OF BDA ENTERING SPRAY FIELD FROM UPWIND
      FLOW=WIND*ASIDE/((81.86*TAIR+22387)*(.03448+HUMID/18))          002080
C        ENTHALPY OF AIR ENTERING SPRAY FIELD,CAL/SEC
      HOUT(1)=FLOW*(.238*TAIR+HUMID*(HVAP+.45*TAIR))
      TSEG(1)=TAIR                                                    002100
      HUM(1)=HUMID                                                    002110
C        CONCENTRATION OF WATER IN AIR
      CWA=HUMID/((81.86*TAIR+22387)*(.03448+HUMID/18))                002120
      EVAPS=0                                                         002130
      SUMTC=0                                                         002140
      DO 1 I=2,11                                                     002150
      TEMP=TSEG(I-1)+273.2                                            002160
C        VISCOSITY OF AIR GM/(CM SEC)
      VIS=2.7936E-6*TEMP**.73617                                      002170
C        DENSITY OF AIR GM/CC
      RHOA=.353/TEMP                                                  002180
C        DIFFUSION COEFFICIENT OF AIR CM**2/SEC
      DIFF=5.8758E-6*TEMP**1.8615
C   .    PRANTL NO
      PR=.93176*TEMP**(-.042784)                                     002200
C        SCHMIDT NO
      SC=2.2705*TEMP**(-.21398)                                      002210
```

Figure B.1   (Continued)

```
C     THERMAL CONDUCTIVITY OF AIR CM/SEC
      AK=3.9273E-7*TEMP**.88315                                    002220
      CON4=AK/R                                                    002230
      CON6=SQRT(2*R*RHOA/VIS)                                      002240
      CON5=DIFF/R                                                  002250
      TDROP=THOT                                                   002260
C     TEMPERATURE AND EVAPORATION OF DROP
      CALL DROP(TSEG(I-1),CWA)                                     002270
C     SENSIBLE HEAT ENTERING SEGMENT FROM DROPS
      HSEG=CON7*(THOT-TDROP)                                       002280
C     EVAPORATION FROM ALL DROPS INTO SEGMENT
      EVAP1=EVAP*CON8                                              002290
C     ENTHALPY LEAVING SEGMENT AND ENTERING NEXT
      HOUT(I)=HOUT(I-1)+HSEG                                       002300
C     HUMIDITY OF SEGMENT
      HUM(I)=HUM(I-1)+EVAP1/FLOW                                   002310
C     AIR TEMPERATURE IN SEGMENT
      TSEG(I)=(HOUT(I)/FLOW-HUM(I)*HVAP)/(.24+.45*HUM(I))          002320
      EVAPS=EVAPS+EVAP1                                            002330
C   CWA = CONCENTRATION OF WATER IN AIR, GM/CC                     002350
      CWA=HUM(I)/((81.86*TSEG(I)+22387)*(.03448+HUM(I)/18))        002340
      SUMTC=SUMTC+TDROP                                            002360
    1 CONTINUE                                                     002370
C     AVERAGE TEMPERATURE OF WATER FALLING TO POND SURFACE
      TWAV=SUMTC/10                                                002380
      RETURN                                                       002390
      END                                                          002400
      SUBROUTINE DROP(TAIR,CINF)                                   002410
      COMMON A,VOL,AM,CON1,CON2,CON3,CON4,CON5,VIS,RHOA,DIFF,      002420
     1 PR ,AK,DT,H,EVAP,NSTEPS,CON6,DTO6,DTO2,TDROP                002430
     1 ,U0,V0,SC                                                   002440
C     CALCULATE HEAT AND MASS TRANSFER FROM A DROP
      EVAP=0                                                       002470
      ICNT=1                                                       002480
C       BEGIN FOURTH ORDER RUNGE-KUTTA INT.OF EQUATIONS            002490
      DO 1 I=1,NSTEPS                                              002500
      CALL FTDROP(ICNT,TDROP,DTD1,DI1,TAIR,CINF)                   002510
      ICNT=ICNT+1                                                  002520
      TDROP1=TDROP+DTO2*DTD1                                       002530
      CALL FTDROP(ICNT,TDROP1,DTD2,DI2,TAIR,CINF)                  002540
      TDROP2=TDROP+DTD2*DTD2                                       002550
      CALL FTDROP(ICNT,TDROP2,DTD3,DI3,TAIR,CINF)                  002560
      ICNT=ICNT+1                                                  002570
      TDROP3=TDROP+DTD3*DT                                         002580
      CALL FTDROP(ICNT,TDROP3,DTD4,DI4,TAIR,CINF)                  002590
      TDROP=TDROP+(DTD1+2*(DTD2+DTD3)+DTD4)*DTO6                   002600
      EVAP=EVAP+(DI1+2*(DI2+DI3)+DI4)*DTO6                         002610
    1 CONTINUE                                                     002620
      RETURN                                                       002630
      END                                                          002640
      SUBROUTINE FTDROP(ICNT,TDRP,DTD,DI,TAIR,CINF)                002650
      COMMON A,VOL,AM,CON1,CON2,CON3,CON4,CON5,VIS,RHOA,DIFF,      002660
     1 PR,AK,DT,H,EVAP,NSTEPS,CON6,DTO6,DTO2,TDROP                 002670
     1 ,U0,V0,SC                                                   002680
C     RATE OF HEAT AND MASS TRANSFER FROM A DROP
      COMMON/RESTOR/ SQV(100)                                      002690
      DATA RG/82.02/                                               002700
      TDK=TDRP+273.2                                               002710
C     VAPOR PRESSURE OF WATER ATM
      P=EXP(71.02499-7381.6477/TDK-9.0993037*ALOG(TDK)            002720
     1 +.0070831556*TDK)                                           002730
```

Figure B.1   (Continued)

```
      SRE=CON6*SQV(ICNT)                                      002740
      HC=CON4*(1+.3*PR**.3333333*SRE)                         002750
      HD=CON5*(1+.3*SC**.3333333*SRE)                         002760
      CDROP=P*18.0/(RG*TDK)                                   002770
C     RATE OF MASS TRANSFER
      DI=CON3*HD*(CDROP-CINF)                                 002780
      DATA HVAP/580.0/
C     RATE OF TEMPERATURE CHANGE
      DTD=-CON1*(DI*HVAP+CON3*HC*(TDRP-TAIR))                 002800
      RETURN                                                  002810
      END                                                     002820
      SUBROUTINE INIT(R,THETA,YO,VELO)
      COMMON A,VOL,AM,CON1,CON2,CON3,CON4,CON5,VIS,RHOA,DIFF, 004170
     1 PR,AK,DT,H,EVAP,NSTEPS,CON6,DTO6,DTO2,TDROP            004180
     1 ,U0,V0,SC                                              004190
      COMMON/RESTOR/SQV(100)                                  004210
      VOL=(3.1415926*4/3)*R**3                                004220
      DATA G/980.0/                                           004230
      DATA HVAP,CP,RHO/597.0,1.0,1.0/                         004240
      A=3.1415926*R**2                                        004250
      CON1=1.0/VOL                                            004260
      CON2=HVAP*12.566371*R**2                                004270
      CON3=12.566371*R**2                                     004280
      V0=VELO*SIN(THETA)                                      004290
      U0=VELO*COS(THETA)                                      004300
      TFALL=V0/G+SQRT((V0/G)**2+2*Y0/G)                       004310
      DT=TFALL/NSTEPS                                         004320
      DTO6=DT/6                                               004330
      DTO2=DT/2                                               004340
      NUM=2*NSTEPS+10                                         004350
      DO 1 I=1,NUM                                            004360
      T=(I-1)*DTO2                                            004370
      V=SQRT(U0**2+(V0-980*T)**2)                             004380
    1 SQV(I)=SQRT(V)                                          004390
      RETURN                                                  004400
      END                                                     004410
      SUBROUTINE FITSPR                                       004440
C     FITS SPRAY EFFICIENCY OF HWS AND LWS MODELS TO REGRESSION
C     EQUATIONS AND COMPARES FITTED RESULTS TO ORIGINAL COMPUTATIONS
      DIMENSION EV(200),YEVAP(200)                            004450
      DIMENSION T(200),TW(200),THOT(200),WIND(200),CH(6),CL(7), 004460
     1 CEH(6),CEL(7),TL(200),TWL(200),THOTL(200),WINDL(200),ETA(200), 004470
     2 ETAL(200),EVAPH(200),EVAPL(200),X(1200),A( 7, 8),P(200),
     3 JJJ( 7),IHLD( 7),YP(200),ETAH(200)
      REWIND 9                                                004510
      REWIND 7                                                004520
      READ(9) NPNTS
      NPL=0                                                   004550
      DO 1 I=1,NPNTS                                          004560
C     READ FROM SCRATCH FILE
      READ(9) TW(I),T(I),THOT(I),WIND(I),HUMID,
     1 TETA,ETAH(I),TEVAP,EVAPH(I)
C     CHECK TO SEE IF LWS MODEL WAS USED                     004600
      IF(TETA.LE.0.0) GOTO 1                                  004610
      NPL=NPL+1                                               004620
      TWL(NPL)=TW(I)                                          004630
      TL(NPL)=T(I)                                            004640
      THOTL(NPL)=THOT(I)                                      004650
      WINDL(NPL)=WIND(I)                                      004660
      ETAL(NPL)=TETA                                          004670
      EVAPL(NPL)=TEVAP                                        004680
C     REVISED SCRATCH FILE ELIMINATING PTS WHERE LWS NOT USED
```

Figure B.1   (Continued)

```
      WRITE(7) TW(I),T(I),THOT(I),WIND(I),HUMID,
     1 TETA,ETAH(I),TEVAP,EVAPH(I)
    1 CONTINUE                                                      004710
      PRINT 101,NPNTS,NPL
  101 FORMAT(10X,'NUMBER OF POINTS GENERATED = ',I5,/
     1 10X,'NUMBER OF POINTS PLOTTED = ',I5)                        004730
C     PUT HWS DATA INTO ARRAY FOR ETA EQN                          004740
      DO 2 I=1,NPNTS                                               004750
      X(I)=T(I)                                                    004760
      I1=I+NPNTS                                                   004770
      I2=I1+NPNTS                                                  004780
      I3=I2+NPNTS                                                  004790
      I4=I3+NPNTS
      X(I1)=TW(I)                                                  004800
      X(I2)=THOT(I)                                                004810
      X(I3)=WIND(I)                                                004820
      X(I4)=SQRT(WIND(I))                                          004840
    2 CONTINUE
C     MULTIPLE REGRESSION ON HWS EFFICIENCY
      CALL SURFIT(X,ETAH,NPNTS,5,7 ,A,WORK,P,JJJ,IHLD,E)
C     SAVE COEFFICIENTS OF EQN FOR ETAH                           004860
      DO 4 I=1,6                                                   004870
    4 CH(I)=A(I,1)                                                 004880
      IF(E.EQ.1.0) WRITE(6,6)
    6 FORMAT(10X,'CONVERGENCE ERROR')                             004900
      WRITE(6,5) (CH(I),I=1,6)
    5 FORMAT(1H1,10X,'FOR HWS EFFICIENCY,CONSTANT AND COEFF OF T,TWET,TH
     1OT,'/,10X,'WIND AND WIND**.5 ARE',/(10X,E15.8))
C     EVAPORATION FOR HWS MODEL                                   004940
C     REGRESSION OF HWS EVAPORATION
      CALL SURFIT(X,EVAPH,NPNTS,5,7 ,A,WORK,P,JJJ,IHLD,E)         004950
      IF(E.EQ.1.0) WRITE(6,6)                                     004970
      DO 7 I=1,6                                                   004980
    7 CEH(I)=A(I,1)
      WRITE(6,8) (CEH(I),I=1,6)
    8 FORMAT(///,10X,'FOR HWS EVAPORATION,CONSTANT AND COEFICIENT OF T,
     1 TWET,THOT,WIND AND WIND**.5 ARE',/(10X,E15.8))
C     SETUP LWS DATA FOR ETAL EQUATION                            005030
      DO 10 I=1,NPL                                                005040
      X(I)=TL(I)                                                   005050
      I1=I+NPL                                                     005060
      I2=I1+NPL                                                    005070
      I3=I2+NPL                                                    005080
      I4=I3+NPL                                                    005090
      I5=I4+NPL
      X(I1)=TL(I)**2
      X(I2)=TL(I)**3
      X(I3)=TWL(I)
      X(I4)=THOTL(I)
      X(I5)=THOTL(I)**2
   10 CONTINUE                                                     005140
C     MULTIPLE REGRESSION FOR LWS EFFICIENCY
      CALL SURFIT(X,ETAL,NPL,6,7 ,A,WORK,P,JJJ,IHLD,E)
      IF(E.EQ.1.0) WRITE(6,6)
C     SAVE COEFF OF EQN FOR ETAL                                  005170
      DO 11 I=1,7
   11 CL(I)=A(I,1)                                                 005190
      WRITE(6,12) (CL(I),I=1,7)
   12 FORMAT(///,10X,'FOR LWS EFFICIENCY,CONSTANT AND COEFF OF T,T**2,
     1 T**3,TWET,THOT AND THOT**2 ARE'/(10X,E15.8))
C     REGRESSION FOR LWS EVAPORATION
      CALL SURFIT(X,EVAPL,NPL,6,7 ,A,WORK,P,JJJ,IHLD,E)
```

Figure B.1   (Continued)

```
        IF(E.EQ.1.0) WRITE(6,6)
        DO 13 I=1,7
     13 CEL(I)=A(I,1)                                           005300
        WRITE(6,14) (CEL(I),I=1,7)
     14 FORMAT(///,10X,'FOR LWS EVAPORATION,CONSTANT AND COEFF OF T,T**2,
      1 T**3,TWET,THOT AND THOT**2 ARE'/(10X,E15.8))
        REWIND 7                                                005360
C       COMPARE REGRESSION TO ORIGINAL
        DO 31 I=1,NPL                                           005370
        READ(7) TW(I),T(I),THOT(I),WIND(I),HUMID,
      1 TETA,ETAH(I),TEVAP,EVAPH(I)
C       CHOOSE HIGHER INPUT EFF
        IF(TETA.GT.ETAH(I)) GOTO 32                             005400
        EV(I)=EVAPH(I)                                          005410
        ETA(I)=ETAH(I)                                          005420
        GOTO 31                                                 005430
     32 ETA(I)=TETA                                             005440
        EV(I)=TEVAP                                             005450
     31 CONTINUE                                                005460
C       PICK HIGHER CORRELATION COEFF                           005470
        DO 33 I=1,NPL                                           005480
        EH=CH(1)+CH(2)*T(I)+CH(3)*TW(I)+CH(4)*THOT(I)+          005490
      1 CH(5)*WIND(I)+CH(6)*SQRT(WIND(I))                       005500
        EL=CL(1)+CL(2)*T(I)+CL(3)*T(I)**2+CL(4)*T(I)**3+
      1 CL(5)*TW(I)+CL(6)*THOT(I)+CL(7)*THOT(I)**2
        IF(EH.GT.EL) GOTO 34
        YP(I)=EL                                                005540
        YEVAP(I)=CEL(1)+CEL(2)*T(I)+CEL(3)*T(I)**2+CEL(4)*T(I)**3  005550
      1 +CEL(5)*TW(I)+CEL(6)*THOT(I)+CEL(7)*THOT(I)**2
        GOTO 33                                                 005580
     34 YP(I)=EH                                                005590
        YEVAP(I)=CEH(1)+CEH(2)*T(I)+CEH(3)*TW(I)+CEH(4)*THOT(I)+  005600
      1 CEH(5)*WIND(I)+CEH(6)*SQRT(WIND(I))
     33 CONTINUE                                                005620
        WRITE(6,81)
     81 FORMAT(1H1,30X,'CORRELATION OF SPRAY EFFICIENCY')
C       PLOT SCATTERGRAMS FOR DATA VS REGRESSION
        CALL SCATTER(ETA,YP,NPL)                                005640
        WRITE(6,82)
     82 FORMAT(1H1,30X,'CORRELATION OF EVAPORATION FRACTION')
        CALL SCATTER(EV,YEVAP,NPL)                              005670
        WRITE(4,201) CH,CL,CEH,CEL
    201 FORMAT(4E15.8)
        STOP
        END                                                    005680
        SUBROUTINE SCATTER(X,Y,NPNTS)                           005690
C       PLOTS SCATTERGRAM OF X ARRAY VS Y ARRAY AND CALCULATES  005700
C       CORRELATION COEFFICIENTS
        DIMENSION ICHAR(11),X(200),Y(200),MA(70,42)            005710
        DATA ICHAR/1H ,1H1,1H2,1H3,1H4,1H5,1H6,1H7,1H8,1H9,1HZ/ 005720
        DO 1 I=1,70                                             005730
        DO 1 J=1,42                                             005740
      1 MA(I,J)=1                                               005740
C       SCALE INPUT                                             005750
        X0=1.0E50                                               005760
        X1=-X0                                                  005770
        SXX=0                                                   005780
        SYY=0                                                   005790
        SXY=0                                                   005800
        XAV=0                                                   005810
        YAV=0                                                   005820
                                                                005830
```

Figure B.1   (Continued)

```
      DO 100 I=1,NPNTS                                              005840
      SXX=X(I)**2+SXX                                              005850
      SYY=Y(I)**2+SYY                                              005860
      SXY=X(I)*Y(I)+SXY                                            005870
      XAV=XAV+X(I)                                                 005880
      YAV=YAV+Y(I)                                                 005890
      IF(X(I).GT.X1) X1=X(I)                                       005900
      IF(Y(I).GT.X1) X1=Y(I)                                       005910
      IF(X(I).LT.X0) X0=X(I)                                       005920
      IF(Y(I).LT.X0)X0=Y(I)                                        005930
  100 CONTINUE                                                     005940
      RANGE=X1-X0                                                  005950
      SXX=NPNTS*SXX-XAV**2                                         005960
      SYY=NPNTS*SYY-YAV**2                                         005970
      SXY=NPNTS*SXY-XAV*YAV                                        005980
      R2=SXY**2/(SXX*SYY)                                          005990
      SERR=SQRT(((SXX*SYY)-SXY**2)/(I*(I-2)*SXX))
      WRITE(6,20) R2,SERR,X0,X1
   20 FORMAT(30X,'CORRELATION COEFF R**2 = ',F8.4,/
     1 30X,'STANDARD ERROR = ',F10.4/,30X,
     1 'MIN AND MAX OF PLOT SCALES = ',2E15.6)
      DO 2 K=1,NPNTS                                               006020
      N=((X(K)-X0)/RANGE)*70+.5                                    006030
      IF(N.GT.70) N=70                                             006040
      IF(N.LT.1) N=1                                               006050
      M=((Y(K)-X0)/RANGE)*42+.5                                    006060
      IF(M.GT.42) M=42                                             006070
      IF(M.LT.1)M=1                                                006080
      MA(N,M)=MA(N,M)+1                                            006090
    2 CONTINUE                                                     006100
      DO 3 N=1,70                                                  006110
      DO 3 M=1,42                                                  006120
      IF(MA(N,M).LT.9) GOTO 4                                      006130
      MA(N,M)=ICHAR(11)                                            006140
      GOTO 3                                                       006150
    4 K=MA(N,M)                                                    006160
      MA(N,M)=ICHAR(K)                                             006170
    3 CONTINUE                                                     006180
      WRITE(6,7)
      DO 5 J=1,42                                                  006200
      J1=42-J+1                                                    006210
    5 WRITE(6,6) (MA(I,J1),I=1,70)
      WRITE(6,7)
    6 FORMAT(26X,1H*,70A1,1H*)
    7 FORMAT(26X,1H*,7(10H1*********))
      WRITE(6,8)
    8 FORMAT(/26X,'PLOTTED CHARACTERS ARE NUMBER OF POINTS FALLING AT TH
     1AT POSITION')
      RETURN
      END                                                         006330
      SUBROUTINE SURFIT(X,Y,N,M,MX,A,WORK,P,JJJ,IHLD,E)           006340
      DIMENSION X(1),Y(1),A(MX,1),WORK(1),P(1),JJJ(1),IHLD(1)     006350
C     MULTIPLE LINEAR REGRESSION ROUTINE
C     R CODELL   AFTER US ARMY MISSILE COMMAND, REDSTONE ARSENAL ALA
      E=0                                                         006360
      L8=M+2                                                      006370
      LV=M+1                                                      006380
      L=1                                                         006390
      JJJ=1                                                       006400
      DO4 I=2,M                                                   006410
      JJJ(I)=N*L+1                                                006420
```

Figure B.1   (Continued)

```
   4 L=L+1                                            006430
     DO 1 I=1,LV                                      006440
     DO 1 J=1,LB                                      006450
   1 A(I,J)=0.                                        006460
     A=N                                              006470
     DO 5 I=1,N                                       006480
   5 P(I)=1.                                          006490
     DO 2 I=1,LV                                      006500
     DO 3 J=1,N                                       006510
   3 A(I,LB)=A(I,LB)+Y(J)*P(J)                        006520
     IF(I.EQ.LV) GOTO 211                             006530
     K=JJJ(I)                                         006540
     DO 2 L=1,N                                       006550
     P(L)=X(K)                                        006560
   2 K=K+1                                            006570
 211 DO 88 I=1,N                                      006580
  88 P(I)=1.                                          006590
     DO 9 I=1,M                                       006600
     LL=I+1                                           006610
     DO 6 J=LL,LV                                     006620
     K=JJJ(J-1)                                       006630
     DO 7 KK=1,N                                      006640
     A(I,J)=A(I,J)+P(KK)*X(K)                         006650
   7 K=K+1                                            006660
   6 A(J,I)=A(I,J)                                    006670
     K=JJJ(I)                                         006680
     DO 9 MM=1,N                                      006690
     P(MM)=X(K)                                       006700
   9 K=K+1                                            006710
     DO 101 I=2,LV                                    006720
     K=JJJ(I-1)                                       006730
     DO 101 KK=1,N                                    006740
     A(I,I)=A(I,I)+X(K)**2                            006750
 101 K=K+1                                            006760
     DO 21 I=1,LV                                     006770
  21 IHLD(I)=I                                        006780
     JJ=LB                                            006790
     DO 55 I=1,LV                                     006800
     KK=LV-I                                          006810
     IF(KK) 10,10,26                                  006820
  26 LL=KK+1                                          006830
     IJJ=1                                            006840
   . L=I                                              006850
     WORK=A                                           006860
     DO 17 II=1,LL                                    006870
     DO 17 J=1,LL                                     006880
     IF(ABS(WORK)-ABS(A(II,J))) 18,17,17             006890
  18 WORK=A(II,J)                                     006900
     L=J+I-1                                          006910
     IJJ=J                                            006920
  17 CONTINUE                                         006930
     IF(IJJ-1)222,222,19                              006940
  19 DO 20 II=1,LV                                    006950
     Z=A(II,1)                                        006960
     A(II,1)=A(II,IJJ)                                006970
  20 A(II,IJJ)=Z                                      006980
     IY=IHLD(I)                                       006990
     IHLD(I)=IHLD(L)                                  007000
     IHLD(L)=IY                                       007010
 222 DO 111 L=1,KK                                    007020
     IF(ABS(A)-ABS(A(L+1,1))) 77,111,111             007030
```

Figure B.1   (Continued)

```
      77 DO 99 J=1,JJ                                          007040
         Z=A(1,J)                                              007050
         A(1,J)=A(L+1,J)                                       007060
      99 A(L+1,J)=Z                                            007070
     111 CONTINUE                                              007080
      10 JJ=JJ+1                                               007090
         IF(A)11,8,11                                          007100
      11 DO 12 J=1,JJ                                          007110
      12 WORK(J)=A(1,J+1)/A                                    007120
         KK=JJ+1                                               007130
         DO 33 K=1,M                                           007140
         DO 33 J=2,KK                                          007150
      33 A(K,J=1)=A(K+1,J)=A(K+1,1)*WORK[J=1)                  007160
         DO 55 J=1,JJ                                          007170
      55 A(LV,J)=WORK(J)                                       007180
         DO 22 I=1,M                                           007190
         L=I+1                                                 007200
         DO 22 J=L,LV                                          007210
         IF(IHLD(I)=IHLD(J)) 22,22,23                          007220
      23 IY=IHLD(I)                                            007230
         IHLD(I)=IHLD(J)                                       007240
         IHLD(J)=IY                                            007250
         Z=A(I,1)                                              007260
         A(I,1)=A(J,1)                                         007270
         A(J,1)=Z                                              007280
      22 CONTINUE                                              007290
      13 RETURN                                                007300
       8 E=1.                                                  007310
         GOTO 13                                               007320
         END                                                   007330
         SUBROUTINE PSY1(DB,WB,PB,DP,PV,W,H,V,RH)              002830
      C  THIS ROUTINE CALCULATES' VAPOR PRESSURE PV, HUMIDITY RATIO W,   002840
      C       ENTHALPY H, VOLUME V, RELATIVE HUMIDITY RH, AND           002850
      C       DEW POINT TEMPERATURE DP\                                 002860
      C       WHEN THE DRY BULB TEMPERATURE DB, WET BULB TEMPERATURE WB, 002870
      C       AND BAROMETRIC PRESSURE PB ARE GIVEN               002880
      C  UNITS' DB, WB, + DP )F>\ PB, + PV )IN OF HG>\ W)= WATER VAPOR    002890
      C       PER = DRY AIR>\ H )BTU/= OF DRY AIR>\ V )FT**3/= OF DRY     002900
      C       AIR\ RH IS A FRACTION, NOT (                        002910
         C(F)=(F-32.0E0)/1.8E0                                 002920
         PVP=PVSF(WB)                                          002930
         WSTAR=0.622*PVP/(PB-PVP)                              002940
         IF (WB.GT.32.0) GO TO 105                             002950
         PV=PVP-5.704E-4*PB*(DB-WB)/1.8                        002960
         GO TO 110                                             002970
     100 PV=PVP                                                002980
         GO TO 110                                             002990
     105 CDB=C(DB)                                             003000
         CWB=C(WB)                                             003010
         HL=597.31+0.4409*CDB=CWB                              003020
         CH=0.2402+0.4409*WSTAR                                003030
         EX=(WSTAR-CH*(CDB=CWB)/HL)/0.622                      003040
         PV=PB*EX/(1.+EX)                                      003050
     110 W=0.622*PV/(PB-PV)                                    003060
         V=0.754*(DB+459.7)*(1.0+7000.0*W/4360.0)/PB           003070
         H=0.24*DB+(1061.0+0.444*DB)*W                         003080
         IF (PV.GT.0.0) GO TO 115                              003090
         PV=0.0                                                003100
         DP=0.0                                                003110
         RH=0.0                                                003120
         RETURN                                                003130
     115 IF (DB.NE.WB) GO TO 120                               003140
```

Figure B.1   (Continued)

```
      DP=DB                                                      003150
      RH=1.0                                                     003160
      RETURN                                                     003170
  120 DP=DPF(PV)                                                 003180
      RH=PV/PVSF(DB)                                             003190
      RETURN                                                     003200
      END                                                        003210
      FUNCTION PVSF(X)                                           003440
      DIMENSION A(6),B(4),P(4)                                   003450
      DATA A/-7.90298,5.02808,-1.3816E-7,11.344,8.1328E-3,-3.49149/  003460
      DATA B/-9.09718,-3.56654,0.876793,0.0060273/              003470
      T=(X+459.688)/1.8                                          003480
      IF (T.LT.273.16) GO TO 100                                 003490
      Z=373.16/T                                                 003500
      P(1)=A(1)*(Z-1.0)                                          003510
      P(2)=A(2)*ALOG10(Z)                                        003520
      Z1=A(4)*(1.0-1.0/Z)                                        003530
      P(3)=A(3)*(10.0**Z1-1.0)                                   003540
      Z1=A(6)*(Z-1.0)                                            003550
      P(4)=A(5)*(10.0**Z1-1.0)                                   003560
      GO TO 105                                                  003570
  100 Z=273.16/T                                                 003580
      P(1)=B(1)*(Z-1.0)                                          003590
      P(2)=B(2)*ALOG10(Z)                                        003600
      P(3)=B(3)*(1.0-1.0/Z)                                      003610
      P(4)=ALOG10(B(4))                                          003620
  105 SUM=0.0                                                    003630
      DO 110 I=1,4                                               003640
  110 SUM=SUM+P(I)                                               003650
      PVSF=29.921*10.0**SUM                                      003660
      RETURN                                                     003670
      END                                                        003680
      FUNCTION DPF(PV)                                           003690
C     THIS ROUTINE CALCULATES DEW-POINT TEMPERATURE FOR A GIVEN  003700
C         VAPOR PRESSURE PV                                      003710
      DP(A,B,C,Y)=A+(B+C*Y)*Y                                    003720
      Y=ALOG(PV)                                                 003730
      IF (PV.GT.0.1836) GO TO 100                                003740
      DPF=DP(71.98,24.873,0.8927,Y)                             003750
      RETURN                                                     003760
  100 DPF=DP(79.047,30.579,1.8893,Y)                            003770
      RETURN                                                     003780
      END                                                        003790
```

Figure B.1   (Continued)

```
      PROGRAM DRIFT(INPUT,OUTPUT,TAPE5=INPUT,TAPE6=OUTPUT)
C                                                                      000110
C     THIS PROGRAM COMPUTES THE DRIFT LOSS FROM A SPRAY POND FOR       000120
C     VARIOUS WIND SPEEDS.  COMPUTATIONS ARE BASED ON A CONSERVATIVE   000130
C     BALLISTIC MODEL OF DROP TRAJECTORIES.                           000140
C     WK NUTTLE AND R CODELL, U.S. NUCLEAR REGULATORY COMMISSION
C     WASHINGTON D.C. 20555
C                                                                      000160
      COMMON WND,VUP,DIA,A,W                                           000170
      REAL GAMMA(2),WIND(16),KX(4),KY(4),DIAM(21)                      000180
      REAL PROPOR(21),DIS(2),RAD(21),XPRIM(21)                         000190
      REAL XI(2,16),YI(2,16),VI(2,16),SPRAY(50,2)
      INTEGER TITLE(60)                                                000210
      DATA GAMMA/0.,180./                                              000220
C     WIND SPEED TABLE
      DATA WIND/0.,2.5,5.,7.5,10.,12.5,15.,17.5,20.,22.5,25.,          000230
     1 30.,35.,40.,45.,50./                                           000240
C     DIAMETER OF DROPS IN TYPICAL SPRAYCO DISTRIBUTION
      DATA DIAM/4000.,3600.,2800.,2290.,2000.,1650.,1340.,1190.,1000.,85000250
     15.,640.,580.,520.,460.,425.,400.,365.,330.,300.,260.,200./      000260
C     FRACTION OF DROPS IN CORRESPONDING DIAMETER RANGE
      DATA PROPOR/.15,.15,.2,.1,.1,.1,.05,.05,.03,.03,.02,.006,.004,.003000270
     1,.002,.001,.001,.001,.001,.0005,.0005/                          000280
C     ASSUMED 50 CM/SEC UPDRAFT IN SPRAY FIELD
      VUP=50                                                           000290
C     XI AND YI ARE COORDINATES OF UPWIND AND DOWNWIND APOGEE FOR
C     EACH WIND SPEED IN TABLE
      DATA XI/235.,-235.,216.,-254.,195.,-270.,173.,-286.,151.,-296.,  000300
     1128.,-306.,104.,-311.,80.,-319.,57.,-327.,32.,-338.,8.,-349.,    000310
     2-38.,-375.,-87.,-403.,-136.,-442.,-185.,-471.,-232.,-512./       000320
      DATA YI/359.,359.,355.,363.,350.,367.,345.,369.,341.,370.,337.,  000330
     1369.,332.,367.,328.,364.,324.,360.,321.,355.,317.,351.,311.,342.,000340
     2305.,333.,299.,325.,294.,318.,290.,311./                        000350
C     VI IS HORIZONTAL DROP VELOCITY AT EACH UPWIND OR DOWNWIND APOGEE
      DATA VI/331.,331.,398.,259.,461.,180.,519.,96.,574.,7.2,626.,    000360
     1-81.,676.,-167.,723.,-246.,769.,-320.,807.,-388.,850.,-451.,932.,000370
     2-566.,1002.,-669.,1069.,-757.,1132.,-844.,1193.,-919./          000380
      NAMELIST/DROPSZ/DIAM,PROPOR
    1 READ(5,520) TITLE                                                000400
  520 FORMAT(80A1)                                                     000410
      IF(TITLE(1).EQ.'S') STOP                                         000420
      WRITE(6,570) TITLE                                               000430
      READ(5,DROPSZ)
  570 FORMAT(1H1,5(/),T20,'TITLE: ',80A1)                              000440
      READ(5,550) NUM                                                  000450
  550 FORMAT(I2)                                                       000460
      WRITE(6,510) NUM                                                 000470
  510 FORMAT(   5(/),T20,'SPRAY GEOMETRY (',I2,' POINTS)'//,T23,       000480
     1'FEET FROM EDGE',T42,'FRAC. OF SPRAYS',/)                        000490
      DO 7 N=1,NUM                                                     000500
      READ(5,560) SPRAY(N,1),SPRAY(N,2)
  560 FORMAT(2F10.0)                                                   000520
    7 WRITE(6,500) SPRAY(N,1),SPRAY(N,2)                               000530
  500 FORMAT(T20,2F15.6,/)
      WRITE(6,540)                                                     000550
  540 FORMAT(1X,5(/),T20,'DRIFT LOSS FRACTION',//,T27,'WIND SPEED',T42,000560
     1'LOSS FRAC.',/)                                                  000570
C     DT IS THE TIMESTEP IN PATHWAY INTEGRATION, SEC
      DATA DT,DTO2,PI/.01,.005,3.1415926/
      DO 20 J=1,16                                                     000600
      WND=WIND(J)*5280./3600.*12.*2.54                                 000610
      DO 2 M=1,21                                                      000620
```

Figure B.2   Listing of program DRIFT

B-15

```
              DIA=DIAM(M)/10000.                                        000630
              A=PI*DIA**2/4                                             000640
              W=PI*DIA**3/6                                             000650
C                                                                       000660
C             INITIALIZE TRAJECTORY CALCULATIONS                        000670
C                                                                       000680
              DO 6 I=1,2                                                000690
              GAM=GAMMA(I)*3.1416/180.                                  000700
              X=XI(I,J)                                                 000710
              Y=YI(I,J)                                                 000720
              VYN=0                                                     000740
              VXN=VI(I,J)*COS(GAM)                                      000750
              DO 3 K=1,1000                                             000770
              CALL FUN(VXN,VYN,KX(1),KY(1))                             000780
              VXN1=VXN-KX(1)*DT                                         000790
              VYN1=VYN-KY(1)*DT                                         000800
              CALL FUN(VXN1,VYN1,KX(2),KY(2))                           000810
              X=X+DTO2*(VXN+VXN1)                                       000820
              Y=Y+DTO2*(VYN+VYN1)                                       000830
              VXN=VXN-DTO2*(KX(1)+KX(2))                                000840
              VYN=VYN-DTO2*(KY(1)+KY(2))                                000850
              DIS(I)=X                                                  000860
              IF(Y.LE.0.) GO TO 6                                       000870
            3 CONTINUE                                                  000880
              RAD(M)=.1                                                 000890
              XPRIM(M)=10000.                                           000900
              GO TO 2                                                   000910
            6 CONTINUE                                                  000920
C                                                                       000930
C             SOLVE FOR RADIUS OF SPRAY DISTRIBUTION AND DISPLACEMENT   000940
C             DOWN WIND                                                 000950
C                                                                       000960
              RAD(M)=(DIS(1)-DIS(2))/2.                                 000970
              XPRIM(M)=(DIS(1)+DIS(2))/((-2.))                          000980
            2 CONTINUE                                                  000990
C                                                                       001000
C             COMPUTE DRIFT LOSS FRACTION                               001010
C                                                                       001020
              DRFTFC=0.                                                 001030
              DO 19 I=1,NUM                                             001040
              XDW=SPRAY(I,1)*12.*2.54                                   001050
              DRFTLS=0.                                                 001060
              DO 18 M=1,21                                              001070
              IF(XDW.GT.(XPRIM(M)+RAD(M)))GO TO 18                      001080
              IF(XDW.GT.(XPRIM(M)-RAD(M)))GO TO 16                      001090
              DRFTLS=DRFTLS+PROPOR(M)                                   001100
              GO TO 18                                                  001110
           16 IF(XDW.GT.XPRIM(M)) GO TO 17                             001120
              DRFTLS=DRFTLS+PROPOR(M)-PROPOR(M)*(ACOS((XPRIM(M)-XDW)/RAD(M))/3.1001130
             14159)                                                    001140
              GO TO 18                                                 001150
           17 DRFTLS=DRFTLS+PROPOR(M)*(ACOS((XDW-XPRIM(M))/RAD(M))/3.14159)001160
           18 CONTINUE                                                 001170
           19 DRFTFC=DRFTLS*SPRAY(I,2)+DRFTFC                          001180
              WRITE(6,530) WIND(J),DRFTFC                              001190
          530 FORMAT(T20,F15.3,F15.8,/)
           20 CONTINUE                                                 001210
              GO TO 1                                                  001220
              END                                                      001230
              SUBROUTINE FUN(VX,VY,DVX,DVY)                            001240
C     VELOCITY COMPONENTS OF DROP                                      001250
              COMMON WND,VUP,DIA,A,W                                   001260
```

Figure B.2   (Continued)

B-16

```
      DATA RHO,VIS/.001204,.0001831/                          001270
C     DROP VELOCITIES WITH RESPECT TO WINDS                   001280
      RVX=VX+WND                                               001290
      RVY=VY-VUP                                               001300
      V=SQRT(RVX**2+RVY**2)                                    001310
      RE=DIA*V*RHO/VIS                                         001320
      IF(RE.GT.2.0) GOTO 11                                    001330
      CD=24/RE                                                 001340
      GOTO 15                                                  001350
   11 IF(RE.GT.500.0) GOTO 12                                  001360
      CD=18.5/RE**.6                                           001370
      GOTO 15                                                  001380
   12 CD=0.44                                                  001390
   15 DRAG=CD*A*RHO*V**2/2                                     001400
      DVX=DRAG*RVX/V/W                                         001410
      DVY=DRAG*RVY/V/W+980.0                                   001420
      RETURN                                                   001430
      END                                                      001440
```

Figure B.2   (Continued)

```
      PROGRAM SPSCAN(INPUT,OUTPUT,TAPE9,TAPE8=/495,TAPE5=INPUT          SPSCAN 2
     1,TAPE6=OUTPUT,PUNCH,TAPE4,DEBUG=OUTPUT)                           SPSCAN 3
C                                                                       SPSCAN 4
C     PROGRAM SPSCAN IS A PROGRAM UNDER DEVELOPMENT BY THE STAFF OF THE SPSCAN 5
C     HYDROLOGIC ENGINEERING SECTION OF THE U.S. NUCLEAR REGULATORY     SPSCAN 6
C     COMMISSION FOR USE IN EVALUATING THE DESIGN BASIS METEOROLOGY OF  SPSCAN 7
C     SMALL SPRAY   PONDS USED AS THE ULTIMATE HEAT SINK OF A NUCLEAR   SPSCAN 8
C     POWER PLANT.  THE PROGRAM USES HISTORICAL WEATHER DATA PROVIDED   SPSCAN 9
C     ON TAPE BY THE NATIONAL WEATHER SERVICE AND A SIMPLIFIED POND     SPSCAN10
C     TEMPERATURE MODEL TO DETERMINE THE PERIOD OF RECORD WHICH WOULD   SPSCAN11
C     RESULT IN EITHER THE   LOWEST COOLING PERFORMANCE OR HIGHEST      SPSCAN12
C     EVAPORATIVE WATER LOSS IN A GIVEN POND.  THE USE OF THE PROGRAM   SPSCAN13
C     AND THE ANALYTICAL TECHNIQUES WHICH IT EMPLOYS ARE FULLY DESCRIBEDSPSCAN14
C     IN LITERATURE AVAILABLE THROUGH THE HYDROLOGIC ENGINEERING        SPSCAN15
C     SECTION.  ALL QUESTIONS AND COMMENTS SHOULD BE ADDRESSED TO       SPSCAN16
C     R. CODELL.                                                        SPSCAN17
C                                                                       SPSCAN18
      REAL LAT1,LAT,YRMODY(3),YRMAX(40,8)                               AUG6   1
      COMMON/COEF/ CEH(6),CEL(7),CH(6),CL(7),FEVAP,FOR,WDRO,NDRIFT,     SPSCAN20
     1 DWDR,FDRIFT(20),HEAT,CON1,CON2,CON3,DTSPRY,DTIME,QSPRAY,CON4,CON5SPSCAN21
     1 ,CEMIN,CEMAX,CMIN,CMAX                                           SPSCAN22
      LAT1=0.                                                           SPSCAN24
      WRITE(6,100)                                                      SPSCAN25
  100 FORMAT(1H1,20(/),10X,'U.S. NUCLEAR REGULATORY COMMISSION- ULTIMATESPSCAN26
     1 HEAT SINK SPRAY   POND METEOROLOGICAL SCANNING MODEL'            SPSCAN27
                                                                        SPSCAN28
      NAMELIST/INPUT/N,A,V,LAT,ISRCH,IPRNT,YRMODY                       SPSCAN29
     1 ,QSPRAY,HEAT,NDRIFT,WDRO,DWDR,FDRIFT                             SPSCAN30
     1 ,CEMIN,CEMAX,CMIN,CMAX                                           SPSCAN31
      HEAT=2.0E8                                                        SPSCAN32
      QSPRAY=50                                                         SPSCAN33
      NDRIFT=3                                                          SPSCAN34
      FDRIFT(1)=0.0                                                     SPSCAN35
      FDRIFT(2)=.00001                                                  SPSCAN36
      FDRIFT(3)=.00002                                                  SPSCAN37
      WDRO=0.0                                                          SPSCAN38
      DWDR=5.0                                                          SPSCAN39
      CMIN=0.1                                                          SPSCAN40
      CMAX=0.8                                                          SPSCAN41
      CEMIN=0.0                                                         SPSCAN42
      CEMAX=0.05                                                        SPSCAN43
      READ(5,555) CH,CL,CEH,CEL                                         SPSCAN44
  555 FORMAT(4E15.8)                                                    SPSCAN45
      DATA N,ISRCH,IPRNT/1,1,0/                                         SPSCAN46
C                                                                       SPSCAN47
C     READ DATA CARD                                                    SPSCAN48
C                                                                       SPSCAN49
    1 READ(5,INPUT)                                                     SPSCAN50
      CON4=QSPRAY*3600                                                  SPSCAN51
      CON3=62.4*3600                                                    SPSCAN52
      CON5=1/(62.4*V)                                                   SPSCAN53
      DTSPRY=HEAT/(QSPRAY*3600*62.4)                                    SPSCAN54
      IF(N.EQ.0) STOP                                                   SPSCAN55
C                                                                       SPSCAN56
C     IF THIS IS THE FIRST DATA CARD OR IF LAT HAS CHANGED, GENERATE A  SPSCAN57
C     NEW INTERMEDIATE FILE.                                            SPSCAN58
C                                                                       SPSCAN59
      IF(ABS(LAT1-LAT).GE..001) CALL SUB1(LAT)                         SPSCAN60
      LAT1=LAT                                                          SPSCAN61
      IF(N.GT.99) GO TO 4                                               SPSCAN62
      IF(V.LT.0.)V=V*(-43560.)                                          SPSCAN63
      IF(A.LT.0.)A=A*(-43560.)                                          SPSCAN64
```

Figure B.3   Listing of program SPSCAN

```
      A1=A/43560.                                                    SPSCAN65
      V1=V/43560.                                                    SPSCAN66
C                                                                    SPSCAN67
C     PRINT POND PARAMETERS.                                         SPSCAN68
C                                                                    SPSCAN69
      WRITE(6,510)N,A,A1,V,V1,ISRCH,IPRNT                            SPSCAN70
  510 FORMAT(5(/),T20,10('*'),' POND NUMBER ',I2,' HAS THE FOLLOWING PARSPSCAN71
     1AMETERS ',25('*'),//,T35,'SURFACE AREA'2X,F12.2,' FT**2 (',F9.2,  SPSCAN72
     2' ACRES)',//,T35,'VOLUME',8X,F12.2,' FT**3 (',F9.2,' ACRE-FT)',//,SPSCAN73
     3T35,'ISRCH = ',I2,T65,'IPRNT = ',I2)                           SPSCAN74
      WRITE(6,550)N                                                  SPSCAN75
  550 FORMAT(5(/),T20,10('*'),' POND NUMBER ',I2,' HAS BEEN MODELLED TO SPSCAN76
     1DETERMINE THE WORST ',13('*'),/,T38,  'PERIODS FOR COOLING AND EVASPSCAN77
     2PORATIVE WATER LOSS',/,1H1)                                    SPSCAN78
      WRITE(6,551)QSPRAY,HEAT,CEMIN,CEMAX,CMIN,CMAX                  SPSCAN79
  551 FORMAT(//,T20,10('*'),'SPRAY PARAMETERS',//,T35,'SPRAY RATE = ',  SPSCAN80
     1F10.2, ' CFS',T35,'BASE HEAT LOAD = ',E12.2,' BTU/HR',/         SPSCAN81
     2,T35,'MINIMUM EVAPORATIVE LOSS FRACTION = ',F10.6,/            SPSCAN82
     3,T35,'MAXIMUM EVAPORATIVE LOSS FRACTION = ',F10.6,/            SPSCAN83
     4,T35,'MINIMUM SPRAY EFFICIENCY = ',F10.4,/,                    SPSCAN84
     5T35,'MAXIMUM SPRAY EFFICIENCY = ',F10.4)                       SPSCAN85
      WRITE(6,552)                                                   SPSCAN86
  552 FORMAT(//,T20,10('*'),'DRIFT LOSS TABLE',//,                   SPSCAN87
     1T30,'WIND SPEED - MPH',T60,'DRIFT LOSS FRACTION')              SPSCAN88
      DO 553 I=1,NDRIFT                                              SPSCAN89
      WINDSP=(I-1)*DWDR+WDR0                                         SPSCAN90
  553 WRITE(6,554)WINDSP,FDRIFT(I)                                   SPSCAN91
  554 FORMAT(/,T35,F10.2,T67,F10.6)                                  SPSCAN92
C                                                                    SPSCAN93
C                                                                    SPSCAN94
C     MODEL TO FIND YEARLY MAXIMUM TEMPERATURES AND 30 DAY EVAPORATIVE SPSCAN94
C     LOSSES.                                                        SPSCAN95
C                                                                    SPSCAN96
      CALL SUB2(A,V,YRMAX)                                           AUG6    2
C     RANK YEARLY MAXIMUM TEMPERATURES AND 30 DAY EVAPORATIVE LOSSES\ SPSCAN98
C     COMPUTE 100 YEAR EXCEEDENCES, SAMPLE MEANS, STANDARD DEVIATIONS. SPSCAN99
C     AND SKEWS.                                                     SPSCA100
C                                                                    SPSCA101
      CALL SUB5(YRMAX)                                               SPSCA102
      IF(ISRCH.LE.0.OR.ISRCH.GE.6) GO TO 1                          SPSCA103
C     PRINT AND/OR PUNCH DAILY METEOROLOGY FOR THE PERIODS OF RECORD SPSCA104
C     PRECEEDING THE HIGHEST ISRCH POND TEMPERATURES. (ISRCH ) 6)    SPSCA105
C                                                                    SPSCA106
      DO 2 I=1,ISRCH                                                 SPSCA107
      DO 3 J=1,3                                                     SPSCA108
      J1=J+1                                                         SPSCA109
    3 YRMODY(J)=YRMAX(I,J1)                                          SPSCA110
      CALL SUB3(YRMODY,IPRNT)                                        AUG6    3
      IF(IPRNT.EQ.1) WRITE(6,520)                                   SPSCA112
  520 FORMAT(1H1)                                                   SPSCA113
    2 CONTINUE                                                      SPSCA127
      GO TO 1                                                       SPSCA128
    4 YRMODY(3)=1.                                                  SPSCA129
C                                                                    SPSCA130
C     CALCULATE AND PRINT MONTHLY AVERAGES OF EACH PARAMETER IN METABL. SPSCA131
C                                                                    SPSCA132
      CALL SUB4(YRMODY,LAT   )                                       SPSCA133
      GO TO 1                                                       SPSCA134
      END                                                           SPSCA135
      SUBROUTINE SUB1(LAT)                                           SUB1    2
C                                                                    SUB1    3
C                                                                    SUB1    4
      REAL METABL(27,10),SRAD(25),LAT                               SUB1    5
      COMMON IDATE(3), IHOUR(6),WINDSP(6),TEMPDB(6),TEMPWB(6),TEMPDP(6),SUB1  6
     1HUMID(6),PRESSR(6),SKY(6)                                      SUB1    7
```

Figure B.3   (Continued)

```
      DATA METABL/270*0./                                          SUB1   8
      DATA SRAD /25*0./                                            SUB1   9
      WRITE(6,520) LAT                                             SUB1  10
  520 FORMAT(5(/),T20,10('*'),' SUBROUTINE SUB1 HAS BEEN CALLED FOR LATISUB1  11
     1TUDE = ',F5.2,' DEG. NORTH ',5('*'),/)                       SUB1  12
C                                                                  SUB1  13
C     POSITION TAPE TO FIRST OF MAY.                               SUB1  14
C                                                                  SUB1  15
      CALL READRC                                                  SUB1  16
      I=(121-IDATE(3))*4-2                                         SUB1  17
      DO 2 J=1,I                                                   SUB1  18
    2 READ(8)                                                      SUB1  19
    3 CALL READRC                                                  SUB1  20
      IF(IHOUR(1).NE.0) GO TO 3                                    SUB1  21
      IF(IDATE(2).LT.5) GO TO 3                                    SUB1  22
C                                                                  SUB1  23
C     READ IN FIRST 6 LINES OF DATA                                SUB1  24
C                                                                  SUB1  25
      DO 4 I=1,6                                                   SUB1  26
      METABL(I,1)=IDATE(1)                                         SUB1  27
      METABL(I,2)=IDATE(2)                                         SUB1  28
      METABL(I,3)=IDATE(3)                                         SUB1  29
      METABL(I,4)=IHOUR(I)                                         SUB1  30
      METABL(I,5)=WINDSP(I)                                        SUB1  31
      METABL(I,6)=TEMPDB(I)                                        SUB1  32
      METABL(I,7)=TEMPDP(I)                                        SUB1  33
      METABL(I,8)=SKY(I)                                           SUB1  34
      METABL(I,9)=TEMPWB(I)                                        SUB1  35
    4 METABL(I,10)=PRESSR(I)                                       SUB1  36
C                                                                  SUB1  37
C     MAKE SURE THAT THE FIRST LINE OF DATA IS COMPLETE.           SUB1  38
C     IF DATA ARE MISSING, SUBSTITUTE FROM THE SECOND OR THIRD LINES SUB1 39
C     IF FIRST THREE LINES ARE BAD, SKIP TO THE NEXT DAY.          SUB1  40
C                                                                  SUB1  41
      INDEX=1                                                      SUB1  42
      IYR=IDATE(1)                                                 SUB1  43
      IMON=IDATE(2)                                                SUB1  44
      IDAY=IDATE(3)                                                SUB1  45
      I=1                                                          SUB1  46
      GO TO 6                                                      SUB1  47
    5 IF(I.EQ.3) GO TO 12                                          SUB1  48
      I=I+1                                                        SUB1  49
      DO 7 J=5,10                                                  SUB1  50
    7 IF(METABL(1,J).GE.999.) METABL(1,J)=METABL(I,J)             SUB1  51
    6 DO 1 J=5,10                                                  SUB1  52
      IF(METABL(1,J).GE.9999.) GO TO 5                            SUB1  53
    1 CONTINUE                                                     SUB1  54
      INDEX=2                                                      SUB1  55
C                                                                  SUB1  56
C     READ IN REST OF FIRST DAY'S DATA.                            SUB1  57
C                                                                  SUB1  58
      DO 8 K=7,19,6                                                SUB1  59
      K5=K+5                                                       SUB1  60
      CALL READRC                                                  SUB1  61
      IK1=I-K+1                                                    SUB1  62
      DO 8 I=K,K5                                                  SUB1  63
      IK1=I-K+1                                                    SUB1  64
      METABL(I,1)=IDATE(1)                                         SUB1  65
      METABL(I,2)=IDATE(2)                                         SUB1  66
      METABL(I,3)=IDATE(3)                                         SUB1  67
      METABL(I,4)=IHOUR(IK1)                                       SUB1  68
      METABL(I,5)=WINDSP(IK1)                                      SUB1  69
      METABL(I,6)=TEMPDB(IK1)                                      SUB1  70
      METABL(I,7)=TEMPDP(IK1)                                      SUB1  71
```

Figure B.3   (Continued).

```
            METABL(I,8)=SKY(IK1)                                          SUB1  72
            METABL(I,9)=TEMPWB(IK1)                                       SUB1  73
        6   METABL(I,10)=PRESSR(IK1)                                      SUB1  74
            CALL READRC                                                   SUB1  75
            DO 9 I=1,3                                                    SUB1  76
            I24=I+24                                                      SUB1  77
            METABL(I24,1)=IDATE(1)                                        SUB1  78
            METABL(I24,2)=IDATE(2)                                        SUB1  79
            METABL(I24,3)=IDATE(3)                                        SUB1  80
            METABL(I24,4)=IHOUR(I)                                        SUB1  81
            METABL(I24,5)=WINDSP(I)                                       SUB1  82
            METABL(I24,6)= TEMPDB(I)                                      SUB1  83
            METABL(I24,7)=TEMPDP(I)                                       SUB1  84
            METABL(I24,8)=SKY(I)                                          SUB1  85
            METABL(I24,9)=TEMPWB(I)                                       SUB1  86
        9   METABL(I24,10)=PRESSR(I)                                      SUB1  87
            METABL(25,4)=24.                                              SUB1  88
      C                                                                   SUB1  89
      C     SEARCH DATA RECORD FOR MISSING DATA AND INTERPOLATE TO        SUB1  90
      C     COMPLETE RECORD.                                              SUB1  91
      C                                                                   SUB1  92
            DO 10 I=1,25                                                  SUB1  93
            DO 10 K=5,10                                                  SUB1  94
            IF (METABL(I,K).LT.9999.) GO TO 10                            SUB1  95
            I1=I+1                                                        SUB1  96
            IF(METABL(I1,K).GE.9999.) GO TO 11                            SUB1  97
            I0=I-1                                                        SUB1  98
            METABL(I,K)=METABL(I1,K)-(METABL(I1,K)-METABL(I0,K))*.5       SUB1  99
            GO TO 10                                                      SUB1 100
       11   I2=I+2                                                        SUB1 101
      C                                                                   SUB1 102
      C     IF THREE OR MORE CONSECUTIVE HOURS OF DATA ARE MISSING, SKIP  SUB1 103
      C     TO THE NEXT DAY.                                              SUB1 104
      C                                                                   SUB1 105
            IF(METABL(I2,K).GE.9999.) GO TO 12                            SUB1 106
            I0=I-1                                                        SUB1 107
            METABL(I,K)=METABL(I2,K)-(METABL(I2,K)-METABL(I0,K))*.6667    SUB1 108
            METABL(I1,K)=METABL(I2,K)-(METABL(I2,K)-METABL(I0,K))*.3333   SUB1 109
       10   CONTINUE                                                      SUB1 110
      C                                                                   SUB1 111
      C     GENERATE SOLAR RADIATION TERM.                                SUB1 112
      C                                                                   SUB1 113
            CALL SOLAR(LAT,IYR,IMON,IDAY,SRAD)                            SUB1 114
      C                                                                   SUB1 115
      C     APPLY CLOUD COVER ADJUSTMENT (AFTER WUNDERLICH) AND READ SOLAR RADSUB1 116
      C     IATION TERM INTO METABL.                                      SUB1 117
      C                                                                   SUB1 118
            DO 13 I=1,25                                                  SUB1 119
       13   METABL(I,8)=SRAD(I)*.94*(1.-.65*METABL(I,8)**2)               SUB1 120
            WRITE ONE DAY'S WEATHER RECORD IN TO INTERMEDIATE STORAGE.    SUB1 121
      C                                                                   SUB1 122
            WRITE(9) METABL                                               SUB1 123
      C                                                                   SUB1 124
      C     IF NEXT DAY IS FIRST OF OCTOBER,SKIP TO NEXT MAY FIRST.       SUB1 125
      C                                                                   SUB1 126
       20   IF(METABL(26,2).LE.9) GO TO 14                                SUB1 127
      C                                                                   SUB1 128
      C     SEPARATE YEARS BY BLANK DATA RECORD.                          SUB1 129
      C                                                                   SUB1 130
            DO 15 I=1,27                                                  SUB1 131
            DO 15 J=1,10                                                  SUB1 132
       15   METABL(I,J)=0.                                                SUB1 133
            WRITE(9) METABL                                               SUB1 134
            DO 16 I=1,847                                                 SUB1 135
```

Figure B.3   (Continued)

```
      READ(8)                                                          SUB1 136
C                                                                      SUB1 137
C        IF END OF RECORD ENCOUNTERED,RETURN TO MAIN PROGRAM.          SUB1 138
C                                                                      SUB1 139
      IF(EOF(8).NE.0) GO TO 17                                         SUB1 140
   16 CONTINUE                                                         SUB1 141
      GO TO 3                                                          SUB1 142
C                                                                      SUB1 143
C        READ IN NEXT DAY'S DATA.                                      SUB1 144
C                                                                      SUB1 145
   14 DO 18 I=1,3                                                      SUB1 146
      I24=I+24                                                         SUB1 147
      DO 18 K=1,10                                                     SUB1 148
   18 METABL(I,K)=METABL(I24,K)                                        SUB1 149
      METABL(1,4)=0.                                                   SUB1 150
      DO 19 I=4,6                                                      SUB1 151
      METABL(I,1)=IDATE(1)                                             SUB1 152
      METABL(I,2)=IDATE(2)                                             SUB1 153
      METABL(I,3)=IDATE(3)                                             SUB1 154
      METABL(I,4)=IHOUR(I)                                             SUB1 155
      METABL(I,5)=WINDSP(I)                                            SUB1 156
      METABL(I,6)=TEMPDB(I)                                            SUB1 157
      METABL(I,7)=TEMPDP(I)                                            SUB1 158
      METABL(I,8)=SKY(I)                                               SUB1 159
      METABL(I,9)=TEMPWB(I)                                            SUB1 160
   19 METABL(I,10)=PRESSR(I)                                           SUB1 161
      INDEX=1                                                          SUB1 162
      IYR=IDATE(1)                                                     SUB1 163
      IMON=IDATE(2)                                                    SUB1 164
      IDAY=IDATE(3)                                                    SUB1 165
      I=1                                                              SUB1 166
      GO TO 6                                                          SUB1 167
C                                                                      SUB1 168
C        WRITE ERROR MESSAGE WHEN DATA ARE SKIPPED                     SUB1 169
C                                                                      SUB1 170
   12 WRITE(6,500) IMON,IDAY,IYR                                       SUB1 171
  500 FORMAT(T35,'DISCONTINUITY IN DATA CAUSED ',I2,'/',I2,'/',I2,' TO BSUB1 172
     1E SKIPPED')                                                      SUB1 173
C                                                                      SUB1 174
C        FLAG RECORD CONTAINING BAD DATA.                             SUB1 175
C                                                                      SUB1 176
      METABL(2,1)=9999.                                                SUB1 177
      WRITE(9) METABL                                                  SUB1 178
      GO TO (3,20),INDEX                                               SUB1 179
   17 REWIND 9                                                         SUB1 180
      REWIND 8                                                         SUB1 181
      RETURN                                                           SUB1 182
      END                                                              SUB1 183
      SUBROUTINE SUB2(A,V,YRMAX)                                       AUG6   4
C     IMPROVED VERSION OF NUTTLE PROGRAM USING 2ND ORDER RK            SUB2   3
C     R CODELL,SEPT 19,1979                                            SUB2   4
C                                                                      SUB2   5
C        MODELS POND TEMPERATURE RESPONSE USING DATA IN INTERMEDIATE   SUB2   6
C        STORAGE.   RETURNS YEARLY MAXIMUM TEMPERATURES AND 30 DAY EVAPOR- SUB2   7
C        ATIVE LOSSES WITH THEIR DATES OF OCCURENCE.                   SUB2   8
C                                                                      SUB2   9
      COMMON/COEF/ CEH(6),CEL(7),CH(6),CL(7),FEVAP,FDR,WDRO,NDRIFT,     SUB2  10
     1 DWDR,FDRIFT(20),HEAT,CON1,CON2,CON3,DTSPRY,DTIME,QSPRAY,CON4,CON5SUB2  11
     1 ,CEMIN,CEMAX,CMIN,CMAX                                          SUB2  12
      REAL ABSMAX(4),METABL(27,10),SRAD(25),TEMPDB(25),                AUG6   5
     1TEMPDP(25),WINDSP(25),KN(4),EV(4),EVAP(30),TEMPMX(5)             SUB2  14
     2,EVPMAX(4),YRMAX(40,8),MAXT                                      AUG6   6
      DIMENSION TEMPWB(25),PRESSR(25)                                  SUB2  16
      DATA DTO2,DTO6,DT/.5,.16666667,1,0/                             SUB2  17
      DO 39 I=1,40                                                     SUB2  18
```

Figure B.3   (Continued)

```
      DO 39 J=1,8                                               SUB2   19
   39 YRMAX(I,J)=0.                                             SUB2   20
      CON1=A/(62.4*24*V)                                        SUB2   21
      CON2=A/(62.4*1040*24)                                     SUB2   22
      LNDX=0                                                    SUB2   23
      MAXT=0.                                                   AUG6    7
      ABSMAX(1)=0.                                              SUB2   29
      EVPMAX(1)=0.                                              SUB2   30
      TEMPMX(1)=0.                                              SUB2   31
      EVTOT=0.                                                  SUB2   32
   10 READ(9) METABL                                           SUB2   33
      IF(EOF(9).NE.0) GO TO 12                                  SUB2   34
      IF(METABL(2,1).GE.9999.) GO TO 10                         SUB2   35
      PONDTP=METABL(1,7)                                        SUB2   36
      DO 30 I=1,30                                              SUB2   37
   30 EVAP(I)=0                                                 SUB2   38
    1 CONTINUE                                                  SUB2   39
      DO 131 J=1,25                                             SUB2   40
      SRAD(J)=METABL(J,8)                                       SUB2   41
      TEMPDB(J)=METABL(J,6)                                     SUB2   42
      TEMPDP(J)=METABL(J,7)                                     SUB2   43
      WINDSP(J)=METABL(J,5)                                     SUB2   44
      TEMPWB(J)=METABL(J,9)                                     SUB2   45
      PRESSR(J)=METABL(J,10)                                    SUB2   46
  131 CONTINUE                                                  SUB2   47
      DO 132 J=1,24                                             SUB2   48
      JP1=J+1                                                   SUB2   49
C                                                               SUB2   50
C     CALCULATION OF POND TEMPERATURE AND EVAPORATIVE WATER LOSS USING SUB2 51
C     THE LINEAR HEAT EXCHANGE EQUATIONS IN A SECOND ORDER RUNGE-KUTTA SUB2 52
C     NUMERICAL INTEGRATION.                                    SUB2   53
C                                                               SUB2   54
      CALL TFUN(PONDTP,TEMPDB(J),WINDSP(J),SRAD(J),TEMPDP(J),   SUB2   55
     1 KN(1),EV(1),TEMPWB(J) ,PRESSR(J))                        SUB2   56
      PTP1=PONDTP+KN(1)*DT                                      SUB2   57
      CALL TFUN(PTP1,TEMPDB(JP1),WINDSP(JP1),SRAD(JP1),TEMPDP(JP1), SUB2 58
     1 KN(2),EV(2),TEMPWB(JP1),PRESSR(JP1))                     SUB2   59
      PONDTP=PONDTP+(KN(1)+KN(2))*DTO2                          SUB2   60
      EVAP(1)=EVAP(1)+(EV(1)+EV(2))*DTO2                        SUB2   61
C                                                               SUB2   62
C     COLLECT MAXIMUM TEMPERATURE                               SUB2   63
C                                                               SUB2   64
      IF(PONDTP.GT.MAXT) MAXT=PONDTP                            AUG6    8
  132 CONTINUE                                                  SUB2   66
C                                                               SUB2   67
C     SEARCH FOR YEARLY MAXIMUM TEMPERATURE AND EVAPORATIVE WATER LOSS. SUB2 68
C                                                               SUB2   69
      DO 33 I=1,30                                              SUB2   70
   33 EVTOT=EVTOT+EVAP(I)                                       SUB2   71
      IF(EVTOT.LT.EVPMAX(1))GO TO 13                            SUB2   72
      EVPMAX(1)=EVTOT                                           SUB2   73
      EVPMAX(2)=METABL(1,1)                                     SUB2   74
      EVPMAX(3)=METABL(1,2)                                     SUB2   75
      EVPMAX(4)=METABL(1,3)                                     SUB2   76
   13 DO 29 I=1,29                                              SUB2   77
      I30=30-I                                                  SUB2   78
      I1=I30+1                                                  SUB2   79
   29 EVAP(I1)=EVAP(I30)                                        SUB2   80
      EVAP(1)=0.                                                SUB2   81
      EVTOT=0.                                                  SUB2   82
      IF(MAXT.LT.ABSMAX(1)) GO TO 8                             AUG6    9
      ABSMAX(1)=MAXT                                            AUG6   10
      ABSMAX(2)=METABL(1,1)                                     SUB2   85
      ABSMAX(3)=METABL(1,2)                                     SUB2   86
      ABSMAX(4)=METABL(1,3)                                     SUB2   87
```

Figure B.3 (Continued)

```
      8 MAXT=0.0                                                          AUG6   11
C                                                                        SUB2   92
C       READ IN NEXT DAY'S DATA.                                         SUB2   93
C                                                                        SUB2   94
     11 READ(9) METABL                                                   SUB2   95
        IF(EOF(9).NE.0.0) GOTO 12                                        SUB2   96
        IF(METABL(1,1).GT.0.) GO TO 14                                   SUB2   97
        LNDX=LNDX+1                                                      SUB2   98
        YRMAX(LNDX,1)=ABSMAX(1)                                          SUB2   99
        YRMAX(LNDX,2)=ABSMAX(2)                                          SUB2  100
        YRMAX(LNDX,3)=ABSMAX(3)                                          SUB2  101
        YRMAX(LNDX,4)=ABSMAX(4)                                          SUB2  102
        YRMAX(LNDX,5)=EVPMAX(1)                                          SUB2  103
        YRMAX(LNDX,6)=EVPMAX(2)                                          SUB2  104
        YRMAX(LNDX,7)=EVPMAX(3)                                          SUB2  105
        YRMAX(LNDX,8)=EVPMAX(4)                                          SUB2  106
        DO 15 I=1,5                                                      SUB2  107
        IF(ABSMAX(1).GE.TEMPMX(1))GO TO 16                               SUB2  108
     15 CONTINUE                                                         SUB2  109
        GO TO 20                                                         SUB2  110
     16 IF(I.GE.5) GO TO 17                                              SUB2  111
        IS=5-I                                                           SUB2  112
        DO 18 J=1,I5                                                     SUB2  113
        L=5-J                                                            SUB2  114
        L1=L+1                                                           SUB2  115
     18 TEMPMX(L1)=TEMPMX(L)                                             SUB2  116
     17 TEMPMX(I)=ABSMAX(1)                                              SUB2  119
     20 ABSMAX(1)=0.                                                     SUB2  122
        EVPMAX(1)=0.                                                     SUB2  123
        MAXT=0.0                                                         AUG6   12
        GO TO 10                                                         SUB2  126
     14 IF(METABL(2,1).LT.9999.) GO TO 1                                 SUB2  127
        MAXT=0.0                                                         AUG6   13
        GO TO 11                                                         SUB2  132
C                                                                        SUB2  133
C       END OF DATA FILE ENCOUNTERED. RETURN TO MAIN PROGRAM.            SUB2  134
C                                                                        SUB2  135
     12 REWIND 9                                                         SUB2  136
        RETURN                                                           SUB2  137
        END                                                             SUB2  138
        SUBROUTINE TFUN(PT,DB,W,SRAD,DP,DT,DE,TW,PINCH)                  TFUN    2
        COMMON/COEF/ CEH(6),CEL(7),CH(6),CL(7),FEVAP,FDR,WDRO,NDRIFT,    TFUN    3
       1 DWOR,FDRIFT(20),HEAT,CON1,CON2,CON3,DTSPRY,DTIME,QSPRAY,CON4,CONSTFUN  4
       1 ,CEMIN,CEMAX,CMIN,CMAX                                          TFUN    5
C       CONVERT PRESSURE TO MM HG                                        TFUN    6
        PAIR=PINCH*25.40                                                 TFUN    7
C       SPRAY HEAT TRANSFER AND WATER LOSS                              TFUN    8
        TSPRAY=PT+DTSPRY                                                 TFUN    9
C       HWS EFFICIENCY                                                   TFUN   10
        ETA=CH(1)+CH(2)*DB+CH(3)*TW+CH(4)*TSPRAY+CH(5)*W+CH(6)*SQRT(W)   TFUN   11
C       LWS EFFICIENCY                                                   TFUN   12
        EL=CL(1)+CL(2)*DB+CL(3)*DB**2+CL(4)*DB**3+CL(5)*TW+              TFUN   13
       1 CL(6)*TSPRAY+CL(7)*TSPRAY**2                                    TFUN   14
        IF(ETA.LT.EL) ETA=EL                                            TFUN   15
        IF(ETA.LT.CMIN) ETA=CMIN                                        TFUN   16
        IF(ETA.GT.CMAX) ETA=CMAX                                        TFUN   17
C       SPRAY HEAT LOSS                                                  TFUN   18
        HSPRAY=HEAT=QSPRAY*CON3*ETA*(TSPRAY-TW)                          TFUN   19
        IF(ETA.EQ.EL) GOTO 3                                            TFUN   20
C       HIGH WIND SPEED EVAPORATION                                     TFUN   21
        FEVAP=CEH(1)+CEH(2)*DB+CEH(3)*TW+CEH(4)*TSPRAY+                  TFUN   22
       1 CEH(5)*W+CEH(6)*SQRT(W)                                         TFUN   23
        GOTO 4                                                          TFUN   24
C       LOW WIND SPEED EVAPORATION                                      TFUN   25
      3 FEVAP=CEL(1)+CEL(2)*DB+CEL(3)*DB**2+CEL(4)*DB**3+CEL(5)*TW        TFUN   26
```

Figure B.3 (Continued)

```
      1 +CEL(6)*TSPRAY+CEL(7)*TSPRAY**2                        TFUN  27
C     DRIFT LOSS                                               TFUN  28
    4 NTBL=(W-WDR0)/DWDR+1                                     TFUN  29
      IF(NTBL.GE.NDRIFT) NTBL=NDRIFT-1                         TFUN  30
      FDR=FDRIFT(NTBL)+((W-WDR0-(NTBL-1)*DWDR)/DWDR)*          TFUN  31
      1 (FDRIFT(NTBL+1)-FDRIFT(NTBL))                          TFUN  32
      IF(FEVAP.LT.CEMIN) FEVAP=CEMIN                           TFUN  33
      IF(FEVAP.GT.CEMAX) FEVAP=CEMAX                           TFUN  34
      ESPRAY=(FDR+FEVAP)*CON4                                  TFUN  35
C     SURFACE HEAT TRANSFER AND EVAPORATION FROM RYAN,1973     TFUN  36
      DTV=(PT+460)/(1-.378*PWAT(PT)/PAIR)-                     TFUN  37
      1 (DB+460)/(1-.378*PWAT(DP)/PAIR)                        TFUN  38
      DTV3=0                                                   TFUN  39
      IF(DTV.LE.0.0) GOTO 1500                                 TFUN  40
      DTV3=DTV**0.33333333                                     TFUN  41
 1500 FU=(22.4*DTV3+14*W)                                      TFUN  42
      HC=0.26*(PT-DB)*FU                                       TFUN  43
      HBR=4.026E-8*(460+PT)**4                                 TFUN  44
      HE=(PWAT(PT)-PWAT(DP))*FU                                TFUN  45
      HAN=1.16E-13*(DB+460)**6*(1-CC**2*.17)                   TFUN  46
C     CONSERVATIVE ASSUMPTION NO CLOUDS                        TFUN  47
      DATA CC/0.0/                                             TFUN  48
      HR=SRAD-HC+HAN-HBR-HE                                    TFUN  49
      OT=HSPRAY*CONS+HR*CON1                                   TFUN  50
      OE=HE*CON2+ESPRAY                                        TFUN  51
      RETURN                                                   TFUN  52
      END                                                      TFUN  53
      FUNCTION PWAT(T)                                         PWAT   2
C     VAPOR PRESSURE OF AIR IN MM HG FOR T IN DEG.F            PWAT   3
      TK=(T-32)/1.8+273.1                                      PWAT   4
      PWAT=760*EXP(71.02499-7381.6677/TK-9.0993037*ALOG(TK)   PWAT   5
      1 +.0070831558*TK)                                       PWAT   6
      RETURN                                                   PWAT   7
      END                                                      PWAT   8
      SUBROUTINE SUB3(YRMODY,IPRNT)                            AUG6  14
C                                                              AUG6  15
C     PRINTS AND/OR PRNCHES DATA FROM INTERMEDIATE             AUG6  16
C     FILE FOR PERIOD OF #NDYS# DAYS BEFORE AND 30             AUG6  17
C     DAYS FOLLOWING YRMODY.                                   AUG6  18
C                                                              AUG6  19
C         IF IPRINT=1,DATA IS PRINTED                          AUG6  20
C         IF IPRINT=-1, DATA IS PUNCHED                        AUG6  21
C         IF IPRINT=0, DATA IS BOTH PRINTED AND PUNCHED        AUG6  22
C                                                              AUG6  23
      REAL YRMODY(3),METABL(27,10),JNDX                        AUG6  24
      INTEGER IDATE(3)                                         AUG6  25
      N=0                                                      AUG6  26
      DATA NDYS/20/                                            AUG6  27
      JNDX=0.                                                  AUG6  28
      IPNCH=0                                                  AUG6  29
      IF(IPRNT.EQ.1) GO TO 40                                  AUG6  30
      IF(IPRNT.EQ.0) IPRNT=1                                   AUG6  31
      IPNCH=1                                                  AUG6  32
   40 CONTINUE                                                 AUG6  33
C                                                              AUG6  34
C     POSITION TAPE9 TO #NDYS# DAYS BEFORE DATE                AUG6  35
C     PROVIDED IN YRMODY.  IF DATA IS NOT AVAILABLE,           AUG6  36
C     POSITION TAPE9 TO FIRST DAY OF DATA IN THE               AUG6  37
C     SAME YEAR AS YRMODY.                                     AUG6  38
C                                                              AUG6  39
      READ(9) METABL                                           AUG6  40
      YR=METABL(1,1)                                           AUG6  41
      REWIND 9                                                 AUG6  42
      IF (YRMODY(1).LE.YR) GO TO 1                             AUG6  43
      N=(YRMODY(1)-YR)*154.                                    AUG6  44
```

Figure B.3 . (Continued).

```
        DO 2 I=1,N                                                   AUG6  45
      2 READ(9) METABL                                               AUG6  46
        N=0                                                          AUG6  47
      1 IF (YRMODY(2).LE.5.)GO TO 3                                  AUG6  48
        N=((YRMODY(2)-5.)*31.)                                       AUG6  49
        IF(YRMODY(2).GT.6.)N=N-1                                     AUG6  50
      3 CONTINUE                                                     AUG6  51
        N=YRMODY(3)+N-NDYS                                           AUG6  52
        IF(N.GT.0)GO TO 4                                            AUG6  53
        NDYS=NDYS+N                                                  AUG6  54
        GO TO 6                                                      AUG6  55
      4 DO 5 I=1,N                                                   AUG6  56
      5 READ(9) METABL                                               AUG6  57
      6 CONTINUE                                                     AUG6  58
        NDYS6=NDYS+30                                                AUG6  59
        N=0                                                          AUG6  60
C                                                                    AUG6  61
C                                                                    AUG6  62
C     GENERATE OUTPUT                                                AUG6  63
C                                                                    AUG6  64
        DO 35 I=1,NDYS6                                              AUG6  64
        READ(9)METABL                                               AUG6  65
        IF(METABL(2,1).GE.9999.)GO TO 35                            AUG6  66
        IF(IPNCH.NE.1) GO TO 41                                      AUG6  67
        IF(I.EQ.1) PUNCH(4,610)NDYS6,METABL(1,2),METABL(1,3),METABL(1,1)  AUG6  68
    610 FORMAT('** APPROXIMATELY ',I2,' DAYS OF MET. DATA FOLLOW. DATA AREAUG6  69
       1PUNCHED 2 HOURS TO A',/,'**** CARD BEGINNING WITH HOUR 0 ON',3F3 AUG6  70
       2.0,'  THE FORMAT FOR THE DATA IS I3,2(',/,'****3F5.1,F6.1,F4.2,F4 AUG6  71
       2.0)WHERE FIELD 1 IS THE CARD NUMBER AND THE FOLLOWING',/,'****VA AUG6  72
       3RIABLE SEQUENCE IS REPEATED:WIND SPEED,DRY BULB,DEWPOINT,SOLAR RA AUG6  73
       5D-',/,'****IATION,CLOUD COVER,AND RELATIVE HUMIDITY.')         AUG6  74
        DO 42 L=1,23,2                                               SUB3  41
        L1=L+1                                                       SUB3  42
        N=N+1                                                        SUB3  43
     42 WRITE(4,590)N,((METABL(J,K),K=5,10),J=L,L1)                  SUB3  44
    590 FORMAT (I3,2(3F5.1,F6.1,F7.2,F7.2))                          SUB3  45
        IF(IPRNT.NE.1) GO TO 35                                      SUB3  46
     41 CONTINUE                                                     SUB3  47
        IDATE(1)=METABL(1,2)                                         SUB3  48
        IDATE(2)=METABL(1,3)                                         SUB3  49
        IDATE(3)=METABL(1,1)                                         SUB3  50
        WRITE(6,500) IDATE                                           SUB3  51
        DO 39 J=1,24                                                 SUB3  52
     39 WRITE(6,520)(METABL(J,K),K=4,10)                             SUB3  53
        WRITE(6,510)                                                 SUB3  54
    500 FORMAT(1H1,5(/),T20,10('*'),' METEOROLOGY FOR '2(I2,'/'),I2,44('*'SUB3  55
       1),///,T25,71('.'),/,T25,',  HOUR  , WIND SP.,DRY BULB ,DEWPOINT ,SUB3  56
       2SOLAR RAD WET BULB ,ATM.PRESS,',/,T25,',',T35,'  (MPH)  , (DEG.F)'SUB3  57
       3 , (DEG.F) ,BTU/FT2/D, (DEG.F) ,  PSIA   ',/,T25,71('.'))       SUB3  58
    510 FORMAT(T25,71('.'))                                          SUB3  59
    520 FORMAT(T25,',',3X,F3.0,3X,',',2X,F4.1,3X,',',2X,F5.1,2X,',',2X,   SUB3  60
       1F5.1,2X,',',2X,F6.1,1X,',',F7.2,2X,',',F7.2,2X,',')          SUB3  61
     35 CONTINUE                                                     SUB3  62
        IF(IPNCH.EQ.1) WRITE(6,600)N                                 SUB3  94
    600 FORMAT(1H1,5(/),T20,10('*'),'NUMBER OF CARDS PUNCHED = ',I3,' ',  SUB3  95
       140('*'))                                                     SUB3  96
        REWIND 9                                                     SUB3  97
        RETURN                                                       SUB3  98
        END                                                          SUB3 103
        SUBROUTINE SUB4(YRMODY,LAT)                                  SUB4   2
C                                                                    SUB4   3
C       PRINTS OUT AVERAGE MONTHLY VALUES FOR METEOROLOGIC PARAMETERS SUB4   4
C       BEGINNING WITH DATE GIVEN IN YRMODY AND ENDING WITH THE LAST SUB4   5
C       DAY ON THE DATA TAPE.                                        SUB4   6
C                                                                    SUB4   7
```

Figure B.3   (Continued)

```
      REAL YRMODY(3),METABL(27,10),LAT                              SUB4    8
      INTEGER IDATE(3),MON(5),MONTH(5)                              SUB4    9
      DATA MON/121,152,182,213,244/                                SUB4   10
      DATA MONTH/'MAY','JUNE','JULY','AUGUST','SEPTEMBER'/          SUB4   11
      INDX=0                                                        SUB4   12
      WINDSP=0.                                                     SUB4   13
      TEMPDP=0.                                                     SUB4   14
      TEMPDB=0.                                                     SUB4   15
      SOLARD=0.                                                     SUB4   16
      IDATE(1)=YRMODY(2)                                            SUB4   17
      PRESSR=0.0                                                    SUB4   18
      TWET=0.0                                                      SUB4   19
      IDATE(2)=YRMODY(3)                                            SUB4   20
      IDATE(3)=YRMODY(1)                                            SUB4   21
      WRITE(6,500) IDATE                                            SUB4   22
  500 FORMAT(   5(/),T20,10('*'),' THE MONTHLY AVERAGE VALUES FROM',SUB4   23
     12(I2,'/'),I2,' TO END OF DATA ',13('*'),//)                  SUB4   24
      WRITE(6,510)                                                  SUB4   25
  510 FORMAT(T30,61('.'),/,T30,'*RMS WIND *DRY BULB *DEWPOINT *  SOLAR  *SUB4   26
     1WET BULB *ATM.PRESS*',/,T30,'*  SPEED  * (DEG.F) * (DEG.F) *RADIATSUB4   27
     2ION* (DEG.F) *  PSIG   *')                                   SUB4   28
      IYR=1900+IDATE(3)                                             SUB4   29
      WRITE(6,520) IYR                                             SUB4   30
  520 FORMAT(T20,I4,T30,61('.'),/,T30,'*',T40,'*',T50,'*',T60,'*',T70,SUB4   31
     1'*',T80,'*',T90,'*')                                        SUB4   32
C                                                                  SUB4   33
C     POSITION TAPE9 TO FIRST DAY OF MONTH PROVIDED IN YRMODY.     SUB4   34
C                                                                  SUB4   35
      READ(9) METABL                                               SUB4   36
      YR=METABL(1,1)                                               SUB4   37
      REWIND 9                                                     SUB4   38
      IF(YRMODY(1).LE.YR) GO TO 1                                  SUB4   39
      N=(YRMODY(1)-YR)*154.+1.                                     SUB4   40
      DO 2 I=1,N                                                   SUB4   41
    2 READ(9)METABL                                                SUB4   42
    1 N=((YRMODY(2)-5.)*31.)                                       SUB4   43
      IF(N.LE.0) GO TO 6                                           SUB4   44
      DO 4 I=1,N                                                   SUB4   45
    4 READ(9) METABL                                               SUB4   46
    6 IF(METABL(1,3).LE.1.) GO TO 5                                SUB4   47
      BACKSPACE 9                                                  SUB4   48
      READ(9)METABL                                                SUB4   49
      GO TO 6                                                      SUB4   50
    5 IF(METABL(2,1).GE.9999.) GO TO 9                             SUB4   51
C                                                                  SUB4   52
C     READ IN ONE MONTH'S DATA                                     SUB4   53
C                                                                  SUB4   54
    8 INDX=INDX+1                                                  SUB4   55
      IDATE(1)=METABL(1,2)                                         SUB4   56
      IDATE(2)=METABL(1,3)                                         SUB4   57
      IDATE(3)=METABL(1,1)                                         SUB4   58
      DAYNUM=MON(IDATE(1)-4)+IDATE(2)-1                            SUB4   59
      IF(MOD(IDATE(3),4).EQ.0) DAYNUM=DAYNUM+1.                    SUB4   60
      DAYLEN=DAYLIT(LAT,DAYNUM)                                    SUB4   61
      DO 7 I=1,24                                                  SUB4   62
      WINDSP=METABL(I,5)**2+WINDSP                                 SUB4   63
      TEMPDB=METABL(I,6)+TEMPDB                                    SUB4   64
      TEMPDP=METABL(I,7)+TEMPDP                                    SUB4   65
      TWET=METABL(I,9)+TWET                                        SUB4   66
      PRESSR=METABL(I,10)+PRESSR                                   SUB4   67
    7 SOLARD=SOLARD+METABL(I,8)/DAYLEN                             SUB4   68
    9 READ (9) METABL                                             SUB4   69
      IF(METABL(1,1).LE.0.) GO TO 11                               SUB4   70
```

Figure B.3   (Continued)

```
      IF(METABL(1,3).LE.1.) GO TO 10                        SUB4   71
      IF(METABL(2,1).GE.9999.) GO TO 9                      SUB4   72
      GO TO 8                                               SUB4   73
   10 DAYS=INDX                                             SUB4   74
C                                                           SUB4   75
C     CALCULATE AND PRINT AVERAGES                          SUB4   76
C                                                           SUB4   77
      INDX=0                                                SUB4   78
      AVGWS=(WINDSP/DAYS/24.)**.5                           SUB4   79
      AVGDP=TEMPDP/DAYS /24.                                SUB4   80
      AVGDB=TEMPDB/DAYS/24.                                 SUB4   81
      AVWET=TWET/DAYS/24                                    SUB4   82
      AVPR=PRESSR/DAYS/24                                   SUB4   83
      AVGSR=SOLARD/DAYS                                     SUB4   84
      I=IDATE(1)-4                                          SUB4   85
      WRITE(6,530) MONTH(I),AVGWS,AVGDB,AVGDP,AVGSR,AVWET,AVPR  SUB4   86
  530 FORMAT(T20,A10,'*',2X,F5.2,2X,'*',2X,F5.2,2X,'*',2X,F5.2,2X,'*',1XSUB4   87
     1,F6.1,2X,'*',F6.2,3X,'*',F6.2,3X,'*',/,T30,'*',T40,'*',T50,  SUB4   88
     3'*',T60,'*',T70,'*',T80,'*',T90,'*')                 SUB4   89
      WINDSP=0.                                             SUB4   90
      TEMPDB=0.                                             SUB4   91
      TWET=0.0                                              SUB4   92
      TEMPDP=0.0                                            SUB4   93
      PRESSR=0.0                                            SUB4   94
      SOLARD=0.                                             SUB4   95
      GO TO 5                                               SUB4   96
   11 DAYS=INDX                                             SUB4   97
C                                                           SUB4   98
C     CALCULATE AND PRINT AVERAGES FOR THE LAST MONTH OF EACH DATA  SUB4   99
C     PERIOD                                                SUB4  100
C                                                           SUB4  101
      INDX=0                                                SUB4  102
      AVGWS=(WINDSP/DAYS/24.)**.5                           SUB4  103
      AVGDP=TEMPDP/DAYS/24.                                 SUB4  104
      AVGDB=TEMPDB/DAYS/24.                                 SUB4  105
      AVPR=PRESSR/DAYS/24                                   SUB4  106
      AVWET=TWET/DAYS/24                                    SUB4  107
      AVGSR=SOLARD/DAYS                                     SUB4  108
      WRITE(6,530) MONTH(5),AVGWS,AVGDB,AVGDP,AVGSR,AVWET,AVPR  SUB4  109
      WINDSP=0.                                             SUB4  110
      TEMPDB=0.                                             SUB4  111
      TEMPDP=0.                                             SUB4  112
      SOLARD=0.                                             SUB4  113
      PRESSR=0.0                                            SUB4  114
      TWET=0.0                                              SUB4  115
      READ (9) METABL                                       SUB4  116
      IF(EOF(9).NE.0) GO TO 12                              SUB4  117
      IYR=1900+METABL(1,1)                                  SUB4  118
      WRITE(6,520) IYR                                      SUB4  119
      IF(METABL(2,1).GE.9999.) GO TO 9                      SUB4  120
      GO TO 8                                               SUB4  121
   12 WRITE(6,540)                                          SUB4  122
  540 FORMAT(T30,61('.'))                                   SUB4  123
      RETURN                                                SUB4  124
      END                                                   SUB4  125
      SUBROUTINE SUB5(YRMAX)                                SUB5    2
C                                                           SUB5    3
C     COMPUTES SAMPLE MEAN, STANDARD DEVIATION,SKEW, AND EXCEEDENCE FOR  SUB5    4
C     YEARLY MAXIMUM TEMPERATURES AND WATER LOSSES GENERATED BY SUB2  SUB5    5
C                                                           SUB5    6
      REAL YRMAX(40,8),JUNK(4),P(40),MT,ME                 SUB5    7
      SUMT=0.                                               SUB5    8
      SUMT2=0.                                              SUB5    9
      SUMT3=0.                                              SUB5   10
```

Figure B.3   (Continued)

```
      SUME=0.                                                          SUB5  11
      SUME2=0.                                                         SUB5  12
      SUME3=0.                                                         SUB5  13
      DO 20 L=1,40                                                     SUB5  14
      IF(YRMAX(L,1).LE.0.) GO TO 21                                    SUB5  15
   20 CONTINUE                                                         SUB5  16
      L=L+1                                                            SUB5  17
   21 L=L-1                                                            SUB5  18
C                                                                      SUB5  19
C     RANK DATA IN ORDER OF DECREASING MAGNITUDE                       SUB5  20
C                                                                      SUB5  21
      DO 1 J=1,5,4                                                     SUB5  22
      DO 1 I=2,L                                                       SUB5  23
      I1=I-1                                                           SUB5  24
      IF(YRMAX(I,J).LE.YRMAX(I1,J)) GO TO 1                            SUB5  25
      DO 2 M=1,4                                                       SUB5  26
      MJ=M+J-1                                                         SUB5  27
    2 JUNK(M)=YRMAX(I,MJ)                                              SUB5  28
      DO 3 M=1,I                                                       SUB5  29
      IF(JUNK(1).GT.YRMAX(M,J)) GO TO 4                                SUB5  30
    3 CONTINUE                                                         SUB5  31
    4 DO 5 K=M,I1                                                      SUB5  32
      KM=I-K+M                                                         SUB5  33
      KM1=KM-1                                                         SUB5  34
      DO 5 L2=1,4                                                      SUB5  35
      LJ=L2+J-1                                                        SUB5  36
    5 YRMAX(KM,LJ)=YRMAX(KM1,LJ)                                       SUB5  37
      DO 6 L2=1,4                                                      SUB5  38
      LJ=L2+J-1                                                        SUB5  39
    6 YRMAX(M,LJ)=JUNK(L2)                                             SUB5  40
    1 CONTINUE                                                         SUB5  41
C                                                                      SUB5  42
C     COMPUTE EXCEEDENCES                                              SUB5  43
C                                                                      SUB5  44
      RL=L                                                             SUB5  45
      P(1)=(1.-(.5)**(1./RL))*100.                                     SUB5  46
      X=2.*(50.-P(1))/(RL-1.)                                          SUB5  47
      DO 7 I=2,L                                                       SUB5  48
      I1=I-1                                                           SUB5  49
    7 P(I)=P(I1)+X                                                     SUB5  50
      DO 22 I=1,L                                                      SUB5  51
      SUMT=SUMT+YRMAX(I,1)                                             SUB5  52
      SUMT2=SUMT2+YRMAX(I,1)**2                                        SUB5  53
      SUMT3=SUMT3+YRMAX(I,1)**3                                        SUB5  54
      SUME=SUME+YRMAX(I,5)                                             SUB5  55
      SUME2=SUME2+YRMAX(I,5)**2                                        SUB5  56
   22 SUME3=SUME3+YRMAX(I,5)**3                                        SUB5  57
      MT=SUMT/RL                                                       SUB5  58
      ST=SQRT((SUMT2-(SUMT**2/RL))/(RL-1.))                            SUB5  59
      GT=(RL**2*SUMT3-3.*RL*SUMT*SUMT2+2.*SUMT**3)/(ST**3*RL*(RL-1.)*  SUB5  60
     1(RL-2.))                                                         SUB5  61
      ME=SUME/RL                                                       SUB5  62
      SE=SQRT((SUME2-(SUME**2/RL))/(RL-1.))                            SUB5  63
      WRITE(6,530)                                                     SUB5  66
  530 FORMAT(/////)                                                    SUB5  67
      WRITE(6,500)                                                     SUB5  68
  500 FORMAT(T20,10('*'),'THE SAMPLE OF YEARLY MAXIMUM POND TEMPERATURESUB5  69
     1 AND 30 DAY  ', 10('*'),/,T31,'EVAPORATIVE LOSSES GENERATED BY THISUB5  70
     2S MODEL IS DESCRIBED BELOW.',////,T28,10('.'),'TEMPERATURE',19('.')SUB5  71
     3,'EVAPORATIVE LOSS',9('.'),/,T28,'*EXCEEDED',15X,  'DATE   *EXCEESUB5  72
     4DED',15X,'DATE   *',/, T28,'*/100 YR* (DEG.F)  *(YR,MO,DY,)*/100SUB5  73
     5 YR*  FT**3  *(YR,MO,DY,)*',/,T28,65('.'))                       SUB5  74
      DO 10 I=1,L                                                      SUB5  75
   10 WRITE(6,510) P(I),(YRMAX(I,J),J=1,4),P(I),(YRMAX(I,K),K=5,8)     SUB5  76
```

Figure B.3   (Continued)

```
  510 FORMAT(T28,'*',1X,F5.2,1X,'*',3X,F5.2,3X,'*',1X,3F3.0,1X,'*',1X,    SUBS  77
     2F5.2,1X,'*',1X,F9.1,1X,'*',1X,3F3.0,1X,'*')                         SUBS  78
        WRITE(6,520) MT,ME,ST,SE                                          STAT1  1
  520 FORMAT(T28,65('.'),//,T26,'MEAN',T40,F5.2,T70,F9.1,//,T17,          SUBS  80
     1 'STANDARD DEV.',T40,F6.3,T70,F10.2)                                STAT1  2
        VART=ST**2                                                        STAT1  3
        VARE=SE**2                                                        STAT1  4
        WRITE(6,600)                                                      STAT1  5
        CALL EXTREM(MT,VART,L)                                            STAT1  6
        WRITE(6,601)                                                      STAT1  7
        CALL EXTREM(ME,VARE,L)                                            STAT1  8
  601 FORMAT(///,35X,'PREDICTED VALUES AND CONFIDENCE LIMITS ON '/,       STAT1  9
     1 35X,'30 DAY EVAPORATION, FT**3'/)                                  STAT1 10
  600 FORMAT(///,35X,'PREDICTED VALUES AND CONFIDENCE LIMITS ON '/,       STAT1 11
     1 35X,'PEAK TEMPERATURE, DEG.F'/)                                    STAT1 12
        RETURN                                                            SUBS  83
        END                                                              SUBS  84
        SUBROUTINE EXTREM(MU,V,N)                                         STAT1 13
C       THIS PROGRAM COMPUTES THE NECESSARY POINTS FOR CONSTRUCTING A     STAT1 14
C       MAXIMUM LIKELIHOOD FREQUENCY CURVE WITH UPPER AND LOWER ERROR BANDSTAT1 15
        REAL EXCD(20),EXHAT(20),TEX(20),SEXC(20),LEX(20),UEX(20)          STAT1 16
        REAL MU                                                           STAT1 17
C       MU= MEAN VALUE                                                    STAT1 18
C       V= VARIENCE                                                       STAT1 19
C       N= SAMPLE SIZE                                                    STAT1 20
C       ALPHA= CONFIDENCE LEVEL FOR ERROR BANDS                           STAT1 21
C              E.G., FOR 5 PER CENT AND 95 PER CENT                       STAT1 22
C              ERROR BANDS ALPHA = .95                                    STAT1 23
        ALPHA=.95                                                         STAT1 24
        NDF=N-1                                                           STAT1 25
        DATA EXCD/.001,.005,.01,.02,.05,.1,.2,.3,.4,.6,.7,.8,             STAT1 26
     1 .9,.95,.98,.99,.995,.999/                                          STAT1 27
        DATA M/18/                                                        STAT1 28
        DO 18 I=1,M                                                       STAT1 29
        PC=EXCD(I)*2.0                                                    STAT1 30
        IF(EXCD(I).GT..5) PC=(1.0-EXCD(I))*2.0                            STAT1 31
        TEX(I)=STUDIN(PC,NDF)                                             STAT1 32
        IF(EXCD(I).GT..5) TEX(I)=-TEX(I)                                  STAT1 33
   18 CONTINUE                                                            STAT1 34
C       COMPUTE EXPECTED VALUE LINE                                       STAT1 35
        DO 21 I=1,M                                                       STAT1 36
        EXHAT(I)=MU+TEX(I)*SQRT(V)                                        STAT1 37
   21   SEXC(I)=SQRT(V*(1.0+.5*TEX(I)**2)/N)                              STAT1 38
C                                                                         STAT1 39
C       COMPUTE UPPER AND LOWER ERROR BANDS                               STAT1 40
C                                                                         STAT1 41
   29   ALPHA=(1.0-ALPHA)*2.0                                             STAT1 42
        TA=STUDIN(ALPHA,NDF)                                              STAT1 43
        DO 31 I=1,M                                                       STAT1 44
        LEX(I)=EXHAT(I)-SEXC(I)*TA                                        STAT1 45
   31   UEX(I)=EXHAT(I)+SEXC(I)*TA                                        STAT1 46
        WRITE(6,200)                                                      STAT1 47
  200 FORMAT(T29,'EXCEEDED',T46,'PREDICTED',T63,'5 PERCENT',              STAT1 48
     1 T81,'95 PERCENT'/,T26,'PER 100 YR',T47,'VALUE',T63,'CONFIDENCE',   STAT1 49
     2 T81,'CONFIDENCE',/)                                                STAT1 50
        DO 60 I=1,M                                                       STAT1 51
        EXCP=EXCD(I)*100.0                                                STAT1 52
   60   WRITE(6,105) EXCP   ,EXHAT(I),LEX(I),UEX(I)                       STAT1 53
  105 FORMAT(1H ,15X,F20.3,3F18.3)                                        STAT1 54
   50   CONTINUE                                                          STAT1 55
        RETURN                                                            STAT1 56
        END                                                               STAT1 57
```

Figure B.3   (Continued)

```
      FUNCTION STUDIN(ALPHA,N)                                    STAT1 58
C                                                                 STAT1 59
C     THIS FUNCTION COMPUTES THE UPPER ALPHA/2 PERCENTILE         STAT1 60
C     POINT FOR A STUDENT'S T DISTRIBUTION WITH N DEGREES OF FREEDOM STAT1 61
C                                                                 STAT1 62
      N1=1                                                        STAT1 63
      N2=N                                                        STAT1 64
      STUDIN=SQRT(FISHIN(ALPHA,N1,N2))                            STAT1 65
      RETURN                                                      STAT1 66
      END                                                         STAT1 67
      FUNCTION FISHIN(ALPHA,N1,N2)                                STAT1 68
C                                                                 STAT1 69
C     THIS FUNCTION COMPUTES THE ALPHA PERCENTILE POINT FOR       STAT1 70
C     FISHER'S F DISTRIBUTION WITH N1 AND N2 DEGREES OF FREEDOM   STAT1 71
C                                                                 STAT1 72
      Y1=N1                                                       STAT1 73
      Y2=N2                                                       STAT1 74
      IF(N1.EQ.1) Y1=2                                            STAT1 75
      IF(N2.EQ.1) Y2=2                                            STAT1 76
      X=TINORM(1.0-ALPHA)                                         STAT1 77
      Y=(X**2-3.0)/6.0                                            STAT1 78
      IC=0                                                        STAT1 79
      Y1=1.0/(Y1-1.0)                                             STAT1 80
      Y2=1.0/(Y2-1.0)                                             STAT1 81
      H=2.0/(Y1+Y2)                                               STAT1 82
      X=X*SQRT(H+Y)/H-(Y1-Y2)*(Y+5.0/6.0-2.0/(3.0*H))            STAT1 83
      X=EXP(2.0*X)                                                STAT1 84
      G=1.0                                                       STAT1 85
      IA1=2                                                       STAT1 86
      IF(MOD(N1,2).EQ.0) GO TO 1                                  STAT1 87
      G=1.7724539                                                 STAT1 88
      IB1=1                                                       STAT1 89
1     IB2=2                                                       STAT1 90
      IF(MOD(N2,2).EQ.0) GO TO 2                                  STAT1 91
      G=G*1.7724539                                               STAT1 92
      IB2=1                                                       STAT1 93
2     IB3=2                                                       STAT1 94
      IF(MOD(N1+N2,2).EQ.0) GO TO 3                               STAT1 95
      G=G/1.7724539                                              STAT1 96
      IB3=1                                                       STAT1 97
3     IF((IB1+IB2).NE.2) G=G*2.0                                 STAT1 98
      IF((N1+N2).LE.3) GO TO 5                                   STAT1 99
      NO=N1+N2-2-IB3                                             STAT1100
      NOP1=NO+1                                                  STAT1101
      DO 4 II=1,NOP1,2                                           STAT1102
      I=II-1                                                     STAT1103
      IF((IB1+I).LE.(N1-2)) G=G*(IB1+I)                          STAT1104
      IF((IB2+I).LE.(N2-2)) G=G*(IB2+I)                          STAT1105
4     G=G/(IB3+I)                                                STAT1106
5     Y2=N2/(N2+N1*X)                                            STAT1107
      Y1=1.0-Y2                                                  STAT1108
      Y=1.0+(G*(1.0-ALPHA-FISH(X,N1,N2)))/SQRT(Y1**N1*Y2**N2)    STAT1109
      FISHIN=X*Y                                                 STAT1110
      IF(Y.LT.0)FISHIN=.5*X                                      STAT1111
      IF(ABS(X/FISHIN-1.0).LT.(.5E-6)) GO TO 7                   STAT1112
      IF(ABS(X-FISHIN).LT.(.5E-6)) GO TO 7                       STAT1113
      IC=IC+1                                                    STAT1114
      IF(IC.GT.100) GO TO 7                                      STAT1115
      X=FISHIN                                                   STAT1116
      GO TO 5                                                    STAT1117
7     RETURN                                                     STAT1118
      END                                                        STAT1119
      FUNCTION TINORM(ALPHA)                                     STAT1120
```

Figure B.3 (Continued)

B-31

```
C                                                                    STAT1121
C     THIS FUNCTION COMPUTES THE ALPHA PERCENTILE FOR THE NORMAL DISTRIBSTAT1122
C                                                                    STAT1123
      DIMENSION A(3),B(3)                                            STAT1124
      DATA A/.010328,.802853,2.515517/, B/.001030R,                 STAT1125
     1 .189269,1.432788/                                            STAT1126
      X=ALPHA                                                       STAT1127
      IF(X) 4,4,1                                                   STAT1128
    1 IF(X-1.0) 2,4,4                                               STAT1129
    2 IF(X.GT.0.5) X=1.0-X                                          STAT1130
      X=SQRT(-2.0*ALOG(X))                                         STAT1131
      TINORN=X-(A(3)+X*(A(2)+X*A(1)))/(1.0+X*(B(3)+X*(B(2)+X*B(1))))STAT1132
      IF(ALPHA.LT..5) TINORN=-TINORN                              STAT1133
    3 TINORM=TINORN                                                STAT1134
      RETURN                                                        STAT1135
    4 TINORN=1.0E38                                                STAT1136
      IF(X.LE.0.0) TINORN=-TINORN                                 STAT1137
      GO TO 3                                                       STAT1138
      END                                                          STAT1139
      FUNCTION FISH(F,N1,N2)                                       STAT1140
C                                                                    STAT1141
C     THIS FUNCTION COMPUTES THE UPPER TAIL AREA OF                 STAT1142
C     FISHER'S F DISTRIBUTION WITH N1 AND N2 DEGREES OF FREEDOM     STAT1143
C                                                                    STAT1144
      LOGICAL E1,E2,E3                                             STAT1145
      E1=.FALSE.                                                   STAT1146
      E2=.FALSE.                                                   STAT1147
      E3=.FALSE.                                                   STAT1148
      IF (MOD(N1,2).EQ.0) E1=.TRUE.                               STAT1149
      IF(MOD(N2,2).EQ.0) E2=.TRUE.                                STAT1150
      X=N2/(N2+N1*F)                                              STAT1151
      IF(.NOT.(E1.OR.E2)) GO TO 5                                 STAT1152
      IF(E1.AND..NOT.E2) GO TO 1                                  STAT1153
      IF(.NOT.E1.AND.E2) GO TO 2                                  STAT1154
      IF(N1.LE.N2) GO TO 1                                        STAT1155
    2 I=N1                                                        STAT1156
      N1=N2                                                       STAT1157
      X=1.0-X                                                     STAT1158
      E3=.TRUE.                                                   STAT1159
    1 Y=1.0-X                                                     STAT1160
      FISH=0.0                                                    STAT1161
      H=SQRT(X**N2)                                               STAT1162
      M=N1/2-1                                                    STAT1163
      MP1=M+1                                                     STAT1164
      DO 3 K=1,MP1                                                STAT1165
      I=K-1                                                       STAT1166
      FISH=FISH+H                                                 STAT1167
    3 H=(H*Y*(N2+2.0*I))/(2.0*K)                                 STAT1168
      IF(E3) GO TO 4                                              STAT1169
      FISH=1.0-FISH                                               STAT1170
      RETURN                                                      STAT1171
    4 I=N1                                                        STAT1172
      N1=N2                                                       STAT1173
      N2=I                                                        STAT1174
      RETURN                                                      STAT1175
    5 Y=1.0-X                                                     STAT1176
      H=.6366197*SQRT(X*Y)                                       STAT1177
      FISH=.6366197*ACOS(SQRT(X))                                STAT1178
      IF(N2.EQ.1) GO TO 8                                         STAT1179
      M=N2-2                                                      STAT1180
      DO 6 I=1,M,2                                                STAT1181
      FISH=FISH+H                                                 STAT1182
```

Figure B.3   (Continued)

B-32

```
      6    H=H+X*(I+1)/(I+2)                                            STAT1183
      *    IF (N1.EQ.1) RETURN                                          STAT1184
           H=H+N2                                                       STAT1185
           M=N1=2                                                       STAT1186
           DO 7 I=1,M,2                                                 STAT1187
           FISH=FISH+H                                                  STAT1188
      7    H=H*Y*(N2+I)/(I+2)                                           STAT1189
           RETURN                                                       STAT1190
           END                                                          STAT1191
           SUBROUTINE READRC                                            READRC 2
      C                                                                 READRC 3
      C    READS WIND SPEED, DRY BULB TEMPERATURE, WET BULB TEMPERATURE, READRC 4
      C    DEW POINT, RELATIVE HUMIDITY, STATION PRESSURE, AND TENTHS OF  READRC 5
      C    CLOUD COVER FROM NATIONAL WEATHER SERVICE DATA TAPES.  WIND SPEED READRC 6
      C    IS RETURNED IN MPH, TEMPERATURE IN DEGREES FARENHEIT, AND PRESSUREREADRC 7
      C    IN MM-HG.  INPUT RECORD IS 495 CHARACTERS LONG.              READRC 8
      C                                                                 READRC 9
           INTEGER JUNK(6,9),ISTAT(2),IWIND(6,4),ITEMP(6,6),IHUMID(6,2), READRC10
          1IPRESS(6,4),ISKY(6,6)                                        READRC11
           COMMON IDATE(3), IHOUR(6),WINDSP(6),TEMPDB(6),TEMPWB(6),TEMPDP(6),READRC12
          1HUMID(6),PRESSR(6),SKY(6)                                    READRC13
           READ(8,500)    ISTAT,IDATE,(IHOUR(I),(JUNK(I,K),K=1,4),       READRC14
          1(IWIND(I,K),K=1,4),(ITEMP(I,K),K=1,6),IHUMID(I,1),IHUMID(I,2), READRC15
          2(IPRESS(I,K),K=1,4),(ISKY(I,K),K=1,6),(JUNK(I,K),K=5,9),I=1,6) READRC16
      500 FORMAT (    I4,I5,3I2,6(I2,1X,I2,A1,1X,I2,A1,I1,A1,4(I2,A1),1X, READRC17
          1I2,A1,I4,A1,I3,A1,1X,6A1,   2(A10),A2,A8,A2,4X))              READRC18
           DO 100 I=1,6                                                 READRC19
           CALL SIGNCK (IWIND(I,3),IWIND(I,4))                          READRC20
           WINDSP(I)=IWIND(I,3)                                         READRC21
           CALL SIGNCK (ITEMP(I,1),ITEMP(I,2))                          READRC22
           WINDSP(I)=WINDSP(I)*1.15078                                  READRC23
           CALL SIGNCK (ITEMP(I,3),ITEMP(I,4))                          READRC24
           CALL SIGNCK (ITEMP(I,5),ITEMP(I,6))                          READRC25
           TEMPDB(I)=ITEMP(I,1)                                         READRC26
           TEMPWB(I)=ITEMP(I,3)                                         READRC27
           TEMPDP(I)=ITEMP(I,5)                                         READRC28
           CALL SIGNCK (IHUMID(I,1),IHUMID(I,2))                        READRC29
           HUMID(I)=IHUMID(I,1)                                         READRC30
           CALL SIGNCK (IPRESS(I,3),IPRESS(I,4))                        READRC31
           PRESSR(I)=IPRESS(I,3)                                        READRC32
           PRESSR(I)=PRESSR(I)*.01                                      READRC33
           ICOVER=0                                                     READRC34
           CALL SIGNCK(ICOVER,ISKY(I,5))                                READRC35
      100 SKY(I)=ICOVER*.1                                              READRC36
           RETURN                                                       READRC37
           END                                                          READRC38
           SUBROUTINE SIGNCK(IFLD,ISGN)                                 SIGNCK 2
      C    THIS SUBROUTINE FURNISHED BY NATIONAL CLIMATIC CENTER, ASHEVILLE SIGNCK 3
      C        WILL TEST ANY PSYCHROMETRIC WITH A SIGN-OVER-UNITS       SIGNCK 4
      C        POSITION READ AS A1 AND THE HIGH ORDER POSITION AS AN    SIGNCK 5
      C        I SPECIFICATION OF PROPER WIDTH                          SIGNCK 6
      C    THE SIGN SHOULD ENTER THE PARAMETER LIST AS ISGN,            SIGNCK 7
      C        THE REMAINING PORTION AS IFLD                            SIGNCK 8
      C    UPON RETURN FROM THE SUBROUTINE THE VALUE OF IFLD WILL BE     SIGNCK 9
      C        AN INTEGER WITH PROPER SIGN                              SIGNCK10
      C    IT WILL BE THE USER'S RESPONSIBILITY TO CONVERT THIS         SIGNCK11
      C        TO   DECIMAL WITH PROPER DECIMAL ALIGNMENT               SIGNCK12
      C    INVALID CONDITION CAUSES IFLD TO BE SET TO 9999              SIGNCK13
           DIMENSION IP(10),MIN(10),NUM(10)                             SIGNCK14
           DIMENSION INUM(10)                                           SIGNCK15
           DATA INUM/'1','2','3','4','5','6','7','8','9','0'/           SIGNCK16
      C    NOTE - SOME COMPUTER SYSTEMS MAY REQUIRE DIFFERENT CHARACTERS AS SIGNCK17
      C    THE LAST CHARACTERS IN ARRAYS IP AND MIN                     SIGNCK18
```

Figure B.3 (Continued)

```
      DATA MIN/'J','K','L','M','N','O','P','Q','R','I'/          SIGNCK19
      DATA IP/'A','B','C','D','E','F','G','H','I',                SIGNCK20
     1 7255555555555555555550/                                   SIGNCK21
      DATA NUM/1,2,3,4,5,6,7,8,9,0/                               SIGNCK22
      DATA IAST/'*'/                                              SIGNCK23
      DATA MINUS/'-'/                                             SIGNCK24
      DATA NULL/' '/                                              SIGNCK25
      IF (ISGN.EQ.NULL.AND.IFLD.NE.0) GO TO 125                   SIGNCK26
      IF (ISGN.EQ.IAST) GO TO 105                                 SIGNCK27
      IF (ISGN.EQ.MINUS) GO TO 110                                SIGNCK28
      DO 100 K=1,10                                               SIGNCK29
      IF (ISGN.EQ.IP(K)) GO TO 115                                SIGNCK30
      IF (ISGN.EQ.MIN(K)) GO TO 120                               SIGNCK31
      IF (ISGN.EQ.INUM(K)) GO TO 115                              SIGNCK32
                                                                  SIGNCK33
  100 CONTINUE                                                    SIGNCK34
  105 IFLD=999999                                                 SIGNCK35
      RETURN                                                      SIGNCK36
  110 IFLD=10                                                     SIGNCK37
      RETURN                                                      SIGNCK38
  125 IFLD=IFLD*10                                                SIGNCK39
      RETURN                                                      SIGNCK40
  115 IFLD=IFLD*10+NUM(K)                                         SIGNCK41
      RETURN                                                      SIGNCK42
  120 IFLD=-(IFLD*10+NUM(K))                                      SIGNCK43
      RETURN                                                      SIGNCK44
      END                                                         SOLAR  2
      SUBROUTINE SOLAR (LAT,YR,MONTH,DAY,SRAD)                    SOLAR  3
C                                                                 SOLAR  4
C     RETURNS INSOLATION IN BTU/FT**2/DAY AT EACH HOUR OF THE DAY. SOLAR 5
C                                                                 SOLAR  5
      INTEGER YR,MONTH,DAY,MONDAT(12)                             SOLAR  6
      REAL LAT,SRAD(25)                                           SOLAR  7
      DATA MONDAT/0,31,59,90,120,151,181,212,243,273,304,334/     SOLAR  8
      LP=MOD(YR,4)                                                SOLAR  9
      IF(LP.NE.0)GO TO 120                                        SOLAR 10
      DO 100 I=3,12                                               SOLAR 11
  100 MONDAT(I)=MONDAT(I)+1                                       SOLAR 12
  120 NUM=MONDAT(MONTH)+DAY                                       SOLAR 13
C                                                                 SOLAR 14
C     FIND TOTAL POSSIBLE DAILY RADIATION AND LENGTH OF DAYLIGHT.  SOLAR 15
C                                                                 SOLAR 16
      TOTRAD=HAMN(LAT,YR,NUM)                                     SOLAR 17
      DAYNUM=NUM                                                  SOLAR 18
      DAYLEN=DAYLIT(LAT,DAYNUM)                                   SOLAR 19
C                                                                 SOLAR 20
C     CALCULATE THE SINUSOIDAL VARIATION IN DAILY RADIATION.      SOLAR 21
C                                                                 SOLAR 22
      T1=.5*DAYLEN                                                SOLAR 23
      A=.5*TOTRAD                                                 SOLAR 24
      DATA W/.2618/                                               SOLAR 25
      ALPHA=1./(1./(A*W)*SIN(W*T1)-T1/A*COS(W*T1))               SOLAR 26
      ALPTO=ALPHA*COS(W*T1)                                      SOLAR 27
      DO 130 I=1,25                                               SOLAR 28
      TO=I-1.                                                     SOLAR 29
      SRAD(I)=0.                                                  SOLAR 30
      TO=ABS(TO-12.)                                             SOLAR 31
C                                                                 SOLAR 32
C     CALCULATE RATE OF INSOLATION FOR EACH HOUR OF DAYLIGHT.     SOLAR 33
C                                                                 SOLAR 34
      IF(TO.LE.T1)SRAD(I)=(ALPHA*COS(W*TO)-ALPTO)*DAYLEN         SOLAR 35
  130 CONTINUE                                                    SOLAR 36
      IF(LP.NE.0) RETURN                                          SOLAR 37
      DO 140 I=3,12                                               SOLAR 38
  140 MONDAT(I)=MONDAT(I)-1                                       SOLAR 39
```

Figure B.3. (Continued)

```
      RETURN                                                          SOLAR 40
      END                                                             SOLAR 41
      FUNCTION DAYLIT(LAT,DAYNUM)                                      DAYLIT 2
C                                                                     DAYLIT 3
C     RETURNS HOURS OF DAYLIGHT GIVEN LATITUDE OF OBSERVATION AND     DAYLIT 4
C     NUMBER OF THE DAY OF THE YEAR.  LATITUDE MUST BE BETWEEN 25 AND DAYLIT 5
C     50 DEGREES NORTH.  THE SOURCE FOR THE LENGTH OF DAYLIGHT INFOR- DAYLIT 6
C     MATION (STORED IN ARRAY 'LENGTH') IS THE SMITHSONIAN METEOROLOG-DAYLIT 7
C     ICAL TABLES.                                                    DAYLIT 8
C                                                                     DAYLIT 9
      REAL LAT,LATBL(6),LENGTH(6,10) ,DAY(10)                         DAYLIT10
      DATA LATBL/25.,30.,35.,40.,45.,50.01/                           DAYLIT11
      DATA DAY/-10.,13.,79.,145.,172.,197.,263.,333.,355.,378./       DAYLIT12
      DATA (LENGTH(1,I),I=1,10)                                       DAYLIT13
     1 /10.58,10.73,12.15,13.50,13.68,13.53,12.17,10.73,10.58,10.73/  DAYLIT14
      DATA (LENGTH(2,I),I=1,10)                                       DAYLIT15
     2/10.20,10.40,12.15,13.83,14.08,13.87,12.17,10.40,10.20,10.40/   DAYLIT16
      DATA (LENGTH(3,I),I=1,10)                                       DAYLIT17
     3/9.80,10.03,12.15,14.23,14.52,14.26,12.20,10.02,9.80,10.03/     DAYLIT18
      DATA (LENGTH(4,I),I=1,10)                                       DAYLIT19
     4/9.33,9.60,12.18,14.67,15.02,14.70,12.22,9.60,9.33,9.60/        DAYLIT20
      DATA (LENGTH(5,I),I=1,10)                                       DAYLIT21
     5/8.75,9.10,12.19,15.28,15.61,15.23,12.23,9.09,8.75,9.10/        DAYLIT22
      DATA (LENGTH(6,I),I=1,10)                                       DAYLIT23
     6/8.07,8.50,12.22,15.83,16.38,15.88,12.28,8.48,8.07,8.50/        DAYLIT24
      DO 100 I=2,10                                                   DAYLIT25
      I1=I-1                                                          DAYLIT26
      IF(DAYNUM.GE.DAY(I1).AND.DAYNUM.LT.DAY(I))GO TO 110             DAYLIT27
  100 CONTINUE                                                        DAYLIT28
  110 DO 120 K=2,6                                                    DAYLIT29
      K1=K-1                                                          DAYLIT30
      IF(LAT.GE.LATBL(K1).AND.LAT.LT.LATBL(K)) GO TO 130             DAYLIT31
  120 CONTINUE                                                        DAYLIT32
C                                                                     DAYLIT33
C     LINEAR INTERPOLATION OF TABLE 'LENGTH'.                         DAYLIT34
C                                                                     DAYLIT35
  130 DELDY=(DAY(I)-DAYNUM)/(DAY(I)-DAY(I1))                          DAYLIT36
      A=LENGTH(K1,I)-(DELDY*(LENGTH(K1,I)-LENGTH(K1,I1)))             DAYLIT37
      B=LENGTH(K,I)-(DELDY*(LENGTH(K,I)-LENGTH(K,I1)))                DAYLIT38
      DAYLIT=B-(LATBL(K)-LAT)/5.*(B-A)                                DAYLIT39
      RETURN                                                          DAYLIT40
      END                                                             DAYLIT41
      FUNCTION HAMN(LAT,YR,MODA)                                      HAMN   2
C     SOLAR RADIATION ON HORIZONTAL SURFACE                           HAMN   3
C         FROM HAMON, WEISS, + WILSON   )100(>                        HAMN   4
C         #MONTHLY WEATHER REVIEW#--PAGE 141--JUNE 1954               HAMN   5
C     PROGRAM AUTHOR--E.C.LONG.  COMPUTER SCIENCES DIVISION--ORNL     HAMN   6
C     UNION CARBIDE NUCLEAR DIVISION.  OAK RIDGE, TENNESSEE           HAMN   7
C **** DAILY RADIATION RETURNED IN BTU'S ****                        HAMN   8
      REAL DATE(16),L25(16),L30(16),L35(16),L40(16),L45(16),L50(16),  HAMN   9
     1     LT(6),LAT,X(3),Y(3),L(96)                                  HAMN  10
      INTEGER IM(12),N(12),YR                                         HAMN  11
      EQUIVALENCE (L(1),L25(1)),(L(17),L30(1)),(L(33),L35(1)),        HAMN  12
     1    (L(49),L40(1)),(L(65),L45(1)),(L(81),L50(1))                HAMN  13
      DATA DATE    /-41.0,-11.0,20.0,51.0,79.0,110.0,140.0,          HAMN  14
     1    171.0,201.0,232.0,263.0,293.0,324.0,354.0,385.0,416.0/     HAMN  15
      DATA L25    /1754.0,1616.0,1794.0,2116.0,2399.0,2611.0,2708.0, HAMN  16
     1    2729.0,2695.0,2571.0,2338.0,2030.0,1754.0,1616.0,          HAMN  17
     2    1794.0,2116.0/                                             HAMN  18
      DATA L30    /1557.0,1390.0,1570.0,1909.0,2266.0,2557.0,2699.0, HAMN  19
     1    2729.0,2662.0,2503.0,2224.0,1873.0,1557.0,1390.0,          HAMN  20
     2    1570.0,1909.0/                                             HAMN  21
      DATA L35    /1338.0,1149.0,1351.0,1723.0,2124.0,2492.0,2680.0, HAMN  22
     1    2729.0,2645.0,2426.0,2064.0,1685.0,1338.0,1149.0,          HAMN  23
```

Figure B.3   (Continued)

```
      2     1351.0,1723.0/                                             HAMN  24
      DATA L40     /1103.0,909.7,1103.0,1514.0,1947.0,2397.0,2655.0,   HAMN  25
      1     2729.0,2603.0,2342.0,1951.0,1479.0,1103.0,909.7,           HAMN  26
      2     1103.0,1514.0/                                             HAMN  27
      DATA L45     /882.7,687.3,881.0,1311.0,1778.0,2289.0,2618.0,     HAMN  28
      1     2729.0,2571.0,2247.0,1769.0,1274.0,882.7,687.3,881.0,1311.0/ HAMN  29
      DATA L50     /682.3,463.3,631.0,1053.0,1568.0,2165.0,2581.0,     HAMN  30
      1     2729.0,2527.0,2136.0,1584.0,1060.0,682.3,463.3,631.7,1053.0/ HAMN  31
      DATA LT   /25.0,30.0,35.0,40.0,45.0,50.0/                        HAMN  32
      DATA   IM   /1,32,60,91,121,152,182,213,244,274,305,335/         HAMN  33
      DATA    N   /31,28,31,30,31,30,31,31,30,31,30,31/                HAMN  34
      DAYC=MODA                                                        HAMN  35
      LEAP=MOD(YR,4)                                                   HAMN  36
      IF (LEAP.NE.0) GO TO 110                                         HAMN  37
      DO 100 I=4,16                                                    HAMN  38
      DATE(I)=DATE(I)+1.0                                              HAMN  39
  100 CONTINUE                                                         HAMN  40
      DO 105 I=2,11                                                    HAMN  41
      IM(I)=IM(I)+1                                                    HAMN  42
      N(I)=N(I)+1                                                      HAMN  43
  105 CONTINUE                                                         HAMN  44
  110 SUM=0.0                                                          HAMN  45
      IF (MODA.GT.0) GO TO 115                                         HAMN  46
C     FOR MODA)0 FIND AVERAGE SOLAR RADIATION FOR MONTH -MODA          HAMN  47
      MO=-MODA                                                         HAMN  48
      I1=IM(MO)                                                        HAMN  49
      ID=N(MO)                                                         HAMN  50
      I2=I1+ID-1                                                       HAMN  51
      DAYS=ID                                                          HAMN  52
      DAY=I1                                                           HAMN  53
      GO TO 120                                                        HAMN  54
C     FOR MODA>0 FIND RADIATION FOR DAY #DAYC#                         HAMN  55
C         DAYC IS EQUIVALENCED TO MODA                                 HAMN  56
  115 I1=1                                                             HAMN  57
      ID=1                                                             HAMN  58
      I2=1                                                             HAMN  59
      DAY=DAYC                                                         HAMN  60
      DAYS=1.0                                                         HAMN  61
  120 DO 180 II=I1,I2                                                  HAMN  62
C     DETERMINE IF DAY IS TABULAR                                      HAMN  63
C         OF IF DAY NOT TABULAR, INDEX OF DAY                          HAMN  64
      MD=0                                                             HAMN  65
      MI=0                                                             HAMN  66
      DO 130 I=2,14                                                    HAMN  67
      DATEI=DATE(I)                                                    HAMN  68
      IF (DAY.NE.DATEI) GO TO 125                                      HAMN  69
      MO=I                                                             HAMN  70
      GO TO 140                                                        HAMN  71
C     MO HAS INDEX I IF DAY=DATE(I)                                    HAMN  72
  125 IF (DAY.GT.DATEI.AND.DAY.LT.DATE(I+1)) GO TO 135                 HAMN  73
  130 CONTINUE                                                         HAMN  74
      GO TO 140                                                        HAMN  75
  135 MI=I                                                             HAMN  76
C     MI=I FOR DATE(I))DAY)DATE(I+1)                                   HAMN  77
C     DETERMINE IF LAT IS TABULAR VALUE                               HAMN  78
  140 IF (MODA.LT.0.AND.II.GT.I1) GO TO 150                            HAMN  79
      ML=0                                                             HAMN  80
      DO 145 I=1,6                                                     HAMN  81
      IF (LAT.NE.LT(I)) GO TO 145                                      HAMN  82
      ML=I                                                             HAMN  83
C     ML=I FOR LAT TABULAR VALUE                                       HAMN  84
      GO TO 150                                                        HAMN  85
  145 CONTINUE                                                         HAMN  86
  150 IF (MO*ML.EQ.0) GO TO 155                                        HAMN  87
```

Figure B.3   (Continued)

```
C         TABULAR DATE + LATITUDE                                    HAMN  88
          J=(ML-1)*16+MD                                            HAMN  89
          HAMN=L(J)                                                 HAMN  90
          GO TO 175                                                 HAMN  91
      155 IF (ML.EQ.0) GO TO 160                                    HAMN  92
C         NON TABULAR DATE + TABULAR LATITUDE                       HAMN  93
          MI1=MI-1                                                  HAMN  94
          J=(ML-1)*16+MI1                                           HAMN  95
          HAMN=YLAG(DAY,DATE(MI1),L(J),4)                           HAMN  96
          GO TO 175                                                 HAMN  97
      160 IF (LAT.LE.32.5) LATF=1                                   HAMN  98
          IF (LAT.GT.32.5.AND.LAT.LE.37.5) LATF=2                   HAMN  99
          IF (LAT.GT.37.5.AND.LAT.LE.42.5) LATF=3                   HAMN 100
          IF (LAT.GT.42.5) LATF=4                                   HAMN 101
          X(1)=LT(LATF)                                             HAMN 102
          X(2)=LT(LATF+1)                                           HAMN 103
          X(3)=LT(LATF+2)                                           HAMN 104
          IF (MD.EQ.0) GO TO 165                                    HAMN 105
C         TABULAR DAY + NON TABULAR LATITUDE                        HAMN 106
          Y(1)=L((LATF-1)*16+MD)                                    HAMN 107
          Y(2)=L(LATF*16+MD)                                        HAMN 108
          Y(3)=L((LATF+1)*16+MD)                                    HAMN 109
          GO TO 170                                                 HAMN 110
C         NON TABULAR DATE + NON TABULAR LATITUDE                   HAMN 111
      165 M1=MI-1                                                   HAMN 112
          Y(1)=YLAG(DAY,DATE(M1),L((LATF-1)*16+M1),4)              HAMN 113
          Y(2)=YLAG(DAY,DATE(M1),L(LATF*16+M1),4)                  HAMN 114
          Y(3)=YLAG(DAY,DATE(M1),L((LATF+1)*16+M1),4)              HAMN 115
      170 HAMN=YLAG(LAT,X,Y,3)                                      HAMN 116
          DAY=DAY+1.0                                               HAMN 117
      175 DAY=DAY+1.0                                               HAMN 118
      180 SUM=SUM+HAMN                                              HAMN 119
          HAMN=AMIN1(2729.0,AMAX1(SUM/DAYS,0.0))                   HAMN 120
          IF (LEAP.NE.0) RETURN                                     HAMN 121
          DO 185 I=4,16                                             HAMN 122
          DATE(I)=DATE(I)-1.0                                       HAMN 123
      185 CONTINUE                                                  HAMN 124
          DO 190 I=2,11                                             HAMN 125
          IM(I)=IM(I)-1                                             HAMN 126
          N(I)=N(I)-1                                               HAMN 127
      190 CONTINUE                                                  HAMN 128
          RETURN                                                    HAMN 129
          END                                                       HAMN 130
          FUNCTION YLAG(XI,X,Y,N)                                   YLAG   2
C         N-POINT LAGRANGIAN INTERPOLATION WHERE I=1,N              YLAG   3
C         SPECIAL VERSION FOR USE WITH FUNCTION #HAMN#              YLAG   4
C         PROGRAM AUTHOR--E.C.LONG.  COMPUTER SCIENCES DIVISION--ORNL YLAG 5
C         UNION CARBIDE NUCLEAR DIVISION.  OAK RIDGE, TENNESSEE     YLAG   6
          DIMENSION X(N),Y(N)                                       YLAG   7
          S=0.0                                                     YLAG   8
          P=1.0                                                     YLAG   9
          DO 110 J=1,N                                              YLAG  10
          P=P*(XI-X(J))                                             YLAG  11
          D=1.0                                                     YLAG  12
          DO 105 I=1,N                                              YLAG  13
          IF (I.NE.J) GO TO 100                                     YLAG  14
          XD=XI                                                     YLAG  15
          GO TO 105                                                 YLAG  16
      100 XD=X(J)                                                   YLAG  17
      105 D=D*(XD-X(I))                                             YLAG  18
      110 S=S+Y(J)/D                                                YLAG  19
          YLAG=S*P                                                  YLAG  20
          RETURN                                                    YLAG  21
          END                                                       YLAG  22
```

Figure B.3 (Continued).

```
         PROGRAM SPRPND(INPUT,OUTPUT,TAPE6=OUTPUT,TAPE8,TAPE5=INPUT)          SPRPND  2
C      PROGRAM TO CALCULATE MAX TEMPERATURE IN A UHS SPRAY-POND               SPRPND  3
C        RICHARD CODELL, U.S.N.R.C. - WASHINGTON D.C. 20555  JULY 1980        SPRPND  4
         DIMENSION TIME(20)                                                   SPRPND  5
         DIMENSION ITITLE(80)                                                 SPRPND  6
         COMMON CH(6),CL(7),CEH(6),CEL(7),NDRIFT,WDRO,DWDR,FDRIFT(20),        SPRPND  7
       1 CEMIN,CEMAX,CMIN,CMAX,VOL,AM,CON1,CON2,CON3,CON4,CON5,CON6,          SPRPND  8
       2 VIS,RHOA,DIFF,AK,H,EVAP,DT06,DT02,TDROP,UO,VO,SC,PRANTL,NSTDR,       SPRPND  9
       3 ATOP(12), ASIDE(12), K1,E,E2,BETA,TSKIP,QBASE,FBASE,M1,M2,BTA,       SPRPND10
       4 BTD,BHS,BW,IMET,BLOW,F1,Q1,TD,TA,HS,W,G(1400,6),HEAT(20),           ITER    1
       5 FLOW(20),TH(20),NMET,NH,A,DTMET,TW,PR,DTDROP                         SPRPND12
       6 ,ASIDEH,HT,WID,ALEN,PB,ISPRAY                                        SPRPND13
         COMMON/SPSW/ TSPRON                                                  SPRPND14
         COMMON/DRPSZ/ R                                                      SPRPND15
         C(Z)=(Z-32)/1.8                                                      SPRPND17
         NAMELIST/HFT/ NH,HEAT,FLOW,TH                                        SPRPND18
         F1=0.0                                                              SPRPND19
         Q1=0.0                                                              SPRPND20
         IMET=0                                                              SPRPND22
         CEMAX=0.1                                                           SPRPND23
         CEMIN=0.                                                            SPRPND24
         CMAX=0.8                                                            SPRPND25
         CMIN=0.2                                                            SPRPND26
         VELO=22.5                                                           JULY30  1
         TA=90.                                                              SPRPND28
         TW=70.                                                              SPRPND29
         TD=60.                                                              SPRPND30
         W=3.                                                                SPRPND31
         HS=1500.                                                            SPRPND32
         PB=29.92                                                            SPRPND33
         THETA=71.0                                                          JULY30  2
         YO=5.0                                                              JULY30  3
         R=.104                                                              SPRPND38
         PHI=90.0                                                            SPRPND39
         NITER=0                                                             ITER    2
         DTITER=5.0                                                          ITER    3
C      NUMBER OF STEPS IN INTEGRATION OF DROP HEAT AND MASS TRANSFER         SPRPND40
         NSTDR=10                                                            SPRPND41
         TZERO=80.0                                                          SPRPND42
         DT=0.2                                                              SPRPND43
         DATA M4,NSTEPS,NPRINT/0,100,10/                                     SPRPND44
         NAMELIST /INLIST/ VZERO,BLOW,A,NH,NSTEPS,NPRINT,DT,TZERO,DTMET      SPRPND45
       1 ,TSKIP,QBASE,FBASE,IMET,        ISPRAY,Q1,F1                        SPRPND46
       2 ,HEAT,FLOW,TH                                                       SPRPND47
       1 ,TA,TW,W,TD,HS,PB,IEVAP,TSPRON                                      SPRPND48
       1 ,NITER,DTITER                                                       ITER    4
         NAMELIST/PARAM/ NDRIFT,WDRO,DWDR,FDRIFT,CEMAX,CEMIN,CMAX,CMIN       SPRPND49
       1 ,VELO,THETA,R,HT,WID,ALEN,YO,PHI,ISPRAY,TA,TD,TW,HS,W,PB           SPRPND50
         READ(5,555) CH,CL,CEH,CEL                                          SPRPND51
   555 FORMAT(4E15.8)                                                       SPRPND52
         READ(5,PARAM)                                                      SPRPND53
C      CONVERT SPRAY PARAMETERS TO METRIC UNITS                             JULY30  4
         VELO=VELO*30.48                                                    JULY30  5
         THETA=THETA*(3.1415926/180.0)                                      JULY30  6
         HT=HT*30.48                                                        JULY30  7
         WID=WID*30.48                                                      JULY30  8
         YO=YO*30.48                                                        JULY30  9
         ALEN=ALEN*30.48                                                    JULY3010
         READ(5,101) NMET                                                   SPRPND54
   101 FORMAT(I5)                                                           SPRPND55
C      READ IN MET TABLE (WIND SP.,DRY BULB, DEW PT,TWET,ATM PRESS)         SPRPND56
C      SKIP FIRST 5 CARDS                                                   JULY3011
```

Figure B.4. Listing of program SPRPND

```
          DO 8 I=1,5                                                     JULY3012
        8 READ(8,9)                                                      JULY3013
        9 FORMAT(1H )                                                    JULY3014
          READ(8,1) (G(I,4),G(I,2),G(I,1),G(I,3),G(I,5),G(I,6),I=1,NMET) SPRPND57
        1 FORMAT(3X,3F5.0,F6.0,2F7.0,3F5.0,F6.0,2F7.0)                   SPRPND58
C     VZERO = VOLUME OF POND FT**3                                       SPRPND59
C     BLOW = BLOWDOWN RATE OUT FT**3/HR                                  SPRPND60
C     A = SURFACE AREA FT**2                                             SPRPND61
C     NSTEPS = NUMBER OF INTEGRATION STEPS                               SPRPND62
C     NPRINT = PRINT EVERY NPRINT STEPS                                  SPRPND63
C     DT = INTEGRATION TIMESTEP, HRS                                     SPRPND64
C     TZERO = INITIAL POND TEMP DEG.F                                    SPRPND65
C     G(I,1)=TD=DEW POINT, DEG.F                                         SPRPND66
C     G(I,2)=TA=DRY BULB DEG.F                                           SPRPND67
C     G(I,3) =HS =  SOLAR RADIATION BTU/(FT**2 DAY)                      SPRPND68
C     G(I,4)= W = WIND SPEED MPH                                         SPRPND69
C     G(I,6)= PR = ATM PRESSURE--INCHES HG                               SPRPND70
C     G(I,5) = TW = WET BULB TEMPERATURE - DEG. F                        SPRPND71
C     QBASE = BASE HEAT LOAD, BTU/HR                                     SPRPND72
C     FBASE = BASE FLOW, FT**3/HR                                        SPRPND73
C     NH = NUMBER OF ENTRIES IN HEAT TABLE                               SPRPND74
C     HEAT = ARRAY OF HEAT INPUTS, BTU/HR                                SPRPND75
C     FLOW = ARRAY OF FLOW RATES, FT**3/HR                               SPRPND76
C     TH = ARRAY OF CORRESPONDING TIMES FOR HEAT AND FLOW ARRAYS         SPRPND77
C     Q1 = HEAT LOAD FOR T LESS THAN TSKIP                               SPRPND78
C     F1 = FLOW FOR T LESS THAN TSKIP                                    SPRPND79
C     (ABOVE 2 USED FOR AMBIENT TEMPERATURE CALCULATION)                 SPRPND80
C     HT = HEIGHT OF SPRAY FIELD, FT                                     JULY3015
C     ALEN = LENGTH OF SPRAY FIELD, FT                                   JULY3016
C     WID = WIDTH OF SPRAY FIELD, FT                                     JULY3017
C     VELO = INITIAL VELOCITY OF DROPS, FT/SEC                           JULY3018
C     THETA = ANGLE OF DROPS WITH RESPECT TO HORIZON, DEGREES            JULY3019
C     YO = HIEGHT OF NOZZLE ABOVE WATER SURFACE, FT                      JULY3020
C     R = THE GEOMETRIC MEAN DROP SIZE, CM                               SPRPND87
C     PB = BAROMETRIC PRESSURE, INCHES HG                                JULY3021
C     CMAX = MAXIMUM ALLOWED SPRAY EFFICIENCY                            SPRPND89
C     CMIN = MINIMUM ALLOWED SPRAY EFFICIENCY                            SPRPND90
C     CEMAX = MAXIMUM ALLOWED EVAPORATION FRACTION                       SPRPND91
C     CEMIN = MINIMUM ALLOWED EVAPORATION FRACTION                       SPRPND92
      BLOW=0                                                             SPRPND93
      OTMET=1                                                            SPRPND94
      QBASE=0                                                            SPRPND96
      FBASE=0                                                            SPRPND97
C**************************************************************************SPRPND98
C     PROGRAM SWITCHES                                                   SPRPND99
C     TSKIP       DELAY START OF HEAT INPUT FROM TABLE TSKIP HOURS        SPRPN100
C                 BEFORE TSKIP HEAT=Q1 AND FLOW=F1                        SPRPN101
C     TSPRON      DELAY SPRAY TURNING ON TSPRON HOURS                     SPRPN102
C                 ALSO ASSUMES FULL POND UNTIL TSPRON HOURS               SPRPN103
C     IEVAP       =1, REGULAR WATER LOSS                                  SPRPN104
C     IEVAP       =0, POND REMAINS FULL - NO WATER LOSS                   SPRPN105
C     ISPRAY      =1, REGRESSION SPRAY MODEL                              SPRPN106
C     ISPRAY      =2, RIGOROUS SPRAY MODEL                                SPRPN107
C     IMET        =0, USE METEOROLOGICAL TABLE AS INPUT                   SPRPN108
C     IMET        =1, FIXED METEOROLOGICAL VARIABLES AS READ IN INLIST    SPRPN109
C**************************************************************************SPRPN110
C     AREA OF SIDE OF SPRAY POND IN HWS MODEL                            SPRPN111
      ASIDEH=HT*ALEN                                                     SPRPN112
      DLEN=ALEN/10                                                       SPRPN113
      DWID=WID/10                                                        SPRPN114
      DO 801 J=1,10                                                      SPRPN115
      I=12-J                                                             SPRPN116
C     TOP AND SIDE AREAS FOR EACH SEGMENT IN LWS MODEL                   JPRPN117
```

Figure B.4   (Continued)

B-39

```
      ATOP(I)=J*DLEN*DWID*J-(J-1)*DLEN*DWID*(J-1)                SPRPN118
      ASIDE(I)=((J-1)*DLEN+(J+1)*DWID)*2*HT                       SPRPN119
  801 CONTINUE                                                    SPRPN120
      ASIDE(1)=(ALEN+WID)*2*HT                                    SPRPN121
      ASIDE(12)=0                                                 SPRPN122
      CALL INIT(R,THETA,YO,VELO)                                  SPRPN123
      READ(5,HFT)                                                 SPRPN124
      DO 4 I=1,NM                                                 SPRPN125
      TIME(I)=TH(I)                                               SPRPN126
      TH(I)=TH(I)+1.0E-20                                         SPRPN127
    4 TH(I)=ALOG(TH(I))                                           SPRPN128
      IF(NH.GT.1) GOTO 710                                        SPRPN129
      FLOW(2)=FLOW(1)                                             SPRPN130
      HEAT(2)=HEAT(1)                                             SPRPN131
      NM=2                                                        SPRPN132
      TH(2)=1.0E8                                                 SPRPN133
  710 CONTINUE                                                    SPRPN134
 6000 CONTINUE                                                    SPRPN135
      ISPRAY=2                                                    JULY3022
      TSPRON=0.0                                                  JULY3023
      TSKIP=0.0                                                   JULY3024
      IEVAP=1                                                     JULY3025
      READ(5,480)ITITLE                                           SPPPN136
  480 FORMAT(80A1)                                                SPRPN137
C     TERMINATE PROGRAM ON A BLANK TITLE CARD                     SPRPN138
      DO 45 I=1,80                                                SPRPN139
      IF(ITITLE(I).NE.1H ) GOTO 46                                SPRPN140
   45 CONTINUE                                                    SPRPN141
      STOP                                                        SPRPN142
   46 CONTINUE                                                    SPRPN143
      READ(5,INLIST)                                              SPRPN144
      Q1S=Q1                                                      ITER   5
      F1S=F1                                                      ITER   6
      WRITE(6,490) ITITLE                                         SPRPN145
  490 FORMAT(1H1,5(/),T20,80A1)                                   SPRPN146
      WRITE(6,200) VELO,THETA,R,HT,WID,ALEN,YO,PHI                SPRPN147
  200 FORMAT(///,20X,'SPRAY FIELD PARAMETERS'/20X,40('*')/        SPRPN148
     1 20X,'INITIAL VELOCITY OF DROPS LEAVING NOZZLE, VELO = ',F10.2,  SPRPN149
     2 ' CM/SEC'/                                                 SPRPN150
     3 20X,'INITIAL ANGLE OF DROPS TO HOR., THETA = ',F10.3,' RADIANS'/ SPRPN151
     4 20X,'GEOMETRIC MEAN RADIUS OF DROPS, R = ',F10.4,' CM'/    SPRPN152
     6 20X,'HEIGHT OF SPRAY FIELD, HT = ',F10.2,' CM'/            SPRPN153
     7 20X,'WIDTH OF SPRAY FIELD, WID = ',F10.1,' CM'/            SPRPN154
     8 20X,'LENGTH OF SPRAY FIELD, ALEN = ',F10.1,' CM'/          SPRPN155
     8 20X,'HEIGHT OF SPRAY NOZZLES ABOVE POND SURFACE, YO = ',F10.1, / SPRPN156
     * 20X,'HEADING OF WIND W.R.T.LONG AXIS, PHI = ',F10.2,' DEGREES'//) SPRPN157
      WRITE(6,500) VZERO,A,BLOW,        NSTEPS,NPRINT,DT,TZERO,   SPRPN158
     1 TSKIP,QBASE,FBASE                                          SPRPN159
  500 FORMAT(///,20X,'POND PARAMETERS'/20X,40('*')/               JULY3026
     120X,'INITIAL POND VOLUME,VZERO = ',F13.1,' CU.FT.'/         JULY3027
     220X,'POND SURFACE AREA,A = ',F13.1,' SQ.FT.'/               JULY3028
     320X,'BLOWDOWN AND LEAKAGE,BLOW = ',F10.2,' CU.FT./HR.'/     JULY3029
     420X,'NUMBER OF INTEGRATION STEPS,NSTEPS = ',I5/             JULY3030
     520X,'PRINT INTERVAL,NPRINT = ',I5/                          JULY3031
     \20X,'INTEGRATION TIMESTEP,DT = ',F10.2,' HOURS'/            JULY3032
     720X,'INITIAL POND TEMPERATURE,TZERO = ',F10.2,' DEG.F'/     JULY3033
     820X,'DELAY FOR HEAT TABLE,TSKIP = ',F10.2,' HRS'/           JULY3034
     920X,'BASE HEAT LOAD ADDED TO TABLE,QBASE = ',F10.2,' HRS'/  JULY3035
     120X,'BASE FLOW RATE ADDED TO TABLE ,FBASE = ',E15.6,' CU.FT./HR.') JULY3036
      WRITE(6,501)                                                JULY3037
  501 FORMAT(////,T43,                                            JULY3038
     635('.'),/,T43,': HEAT IN  : TIME FROM : FLOW IN  :',/,T43,':  BTU/ SPRPN166
     7HR  :   START    : FT**3/HR :',/,T43,35('.'))               SPRPN167
```

Figure B.4   (Continued)

```
      DO 2 I=1,NH                                              SPRPN168
    2 WRITE(6,510)HEAT(I),TIME(I),FLOW(I)                      SPRPN169
  510 FORMAT(T43,'!',    E9.3,1X,'!',2X,F7.2,2X,'!',    E9.3,1X,'!')  SPRPN170
      WRITE(6,524) Q1,F1                                       SPRPN171
  524 FORMAT(/T30,'FOR TIME LESS THAN TSKIP'/T30,'Q1 = ',E12.3, SPRPN172
     1 ' BTU/HR'/T30,'F1 = ',E12.3,' FT**3/HR')               SPRPN173
      IF(IMET.EQ.0) WRITE(6,47)                                SPRPN174
   47 FORMAT(/20X,'METEOROLOGICAL TABLE USED AS INPUT'/)       SPRPN175
      IF(IMET.EQ.1) WRITE(6,48)                                SPRPN176
   48 FORMAT(/20X,'FIXED METEOROLOGICAL VALUES USED AS INPUT'/) SPRPN177
      IF(IMET.EQ.1)WRITE(6,61)TA,TW,W,TD,HS,PB                 JULY3039
   61 FORMAT(/20X,'DRY BULB TEMPERATURE,TA = ',F10.2,' DEG. F'/ JULY3040
     120X,'WET BULB TEMPERATURE,TW = ',F10.2,' DEG. F'/        JULY3041
     220X,'WIND SPEED,W = ',F10.2,'MPH'/                       JULY3042
     320X,'DEW POINT TEMPERATURE,TD = ',F10.2,' DEG. F'/       JULY3043
     420X,'SOLAR RADIATION,HS = ',F10.2,' BTU/SQ.FT./DAY'/     JULY3044
     520X,'BAROMETRIC PRESSURE,PB = ',F10.2,' IN.HG.')         JULY3045
      IF(ISPRAY.EQ.2) WRITE(6,49)                              SPRPN178
   49 FORMAT(/20X,'RIGOROUS SPRAY MODEL CHOSEN'/)              SPRPN179
      IF(ISPRAY.NE.2) WRITE(6,50)                              SPRPN180
   50 FORMAT(/20X,'REGRESSION EQUATIONS USED FOR SPRAY MODEL'/) SPRPN181
      WRITE(6,53) TSPRON                                       SPRPN182
   53 FORMAT(/20X,'SPRAYS WILL BE DELAYED',F10.2,1X,'HOURS',/) SPRPN183
      WRITE(6,520)                                             SPRPN184
  520 FORMAT(T43,35('.'),5(/),T41,13('*'),' MODEL RESULTS ',13('*'),////, SPRPN185
     1T38,'..TIME.......TEMPERATURE (F).........VOLUME....',/,T38,': HR SPRPN186
     2                        :   FT**3   :',/,T38,46('.'))    SPRPN187
 6003 CONTINUE                                                 ITER   7
      TS=0                                                     ITER   8
      M4=0                                                     ITER   9
      F1=F1S                                                   ITER  10
      Q1=Q1S                                                   ITER  11
      T5=0                                                     SPRPN189
      M1=1                                                     SPRPN190
      M2=1                                                     SPRPN191
      X=.001                                                   SPRPN192
      T=TZERO                                                  SPRPN193
      V=VZERO                                                  SPRPN194
      VMIN=0.1*VZERO                                           ITER  12
    C  BEGIN NUMERICAL INTEGRATIONS                            SPRPN195
      DO 6 M=1,NSTEPS                                          SPRPN196
    C  MIXED TANK SOLUTIONS                                    SPRPN197
      CALL MIXED(F2,F3,T,V,X)                                  SPRPN198
    C  FORCE FULL POND IF IEVAP≠0                              SPRPN199
      IF(IEVAP.EQ.0.OR.X.LT.TSPRON) F3=0.0                     SPRPN200
      CALL MIXED(F7,F8,T+DT*F2,V+DT*F3,X+DT)                   SPRPN201
      IF(IEVAP.EQ.0.OR.X.LT.TSPRON) F8=0.0                     SPRPN202
      T=T+DT*(F2+F7)/2                                         SPRPN203
      V=V+DT*(F3+F8)/2                                         SPRPN204
      IF(V.LT.VMIN) V=VMIN                                     ITER  13
    C  FIND MAX TEMPERATURE FOR MIXED MODEL                    SPRPN205
      IF(T.LT.T5) GOTO 63                                      SPRPN206
      T5=T                                                     SPRPN207
      TIMEM=X                                                  SPRPN208
   63 CONTINUE                                                 SPRPN209
      M4=M4+1                                                  SPRPN210
      X=X+DT                                                   SPRPN211
      IF(NPRINT.GT.M4) GOTO 6                                  SPRPN212
      M4=0                                                     SPRPN213
      WRITE(6,51) X,T,V                                        SPRPN214
   51 FORMAT(T35,F10.2,T53,F10.2,T70,E15.8)                    SPRPN215
    6 CONTINUE                                                 SPRPN216
      IF(NITER.EQ.0) WRITE(6,566)                              ITER  14
```

Figure B.4   (Continued)

```
  566 FORMAT('1H0)                                                   ITER   15
      WRITE(6,55) TSKIP,TS,TIMEM                                     ITER   16
   55 FORMAT ( T5,'TSKIP = ',F8.1,' HOURS',5X,'MAX MODELED TEMPERATURE ITER  17
     1= ',F8.2,' AT',F8.2,' HOURS')                                  ITER   18
      IF(NITER.LE.0) GOTO 6001                                       ITER   19
      TSPRON=TSPRON+DTITER                                           ITER   20
      TSKIP=TSKIP+DTITER                                             ITER   21
      NITER=NITER-1                                                  ITER   22
      GOTO 6003                                                      ITER   23
 6001 CONTINUE                                                       ITER   24
      GOTO 6000                                                      SPRPN220
      END                                                            SPRPN221
      SUBROUTINE MIXED(FA,FB,T,V,X)                                  MIXED   2
C     MIXED TANK MODEL                                               MIXED   3
      COMMON CH(6),CL(7),CEH(6),CEL(7),NDRIFT,WDRO,DWDR,FDRIFT(20),  MIXED   4
     1 CEMIN,CEMAX,CMIN,CMAX,VOL,AM,CON1,CON2,CON3,CON4,CON5,CON6,   MIXED   5
     2 VIS,RHOA,DIFF,AK,H,EVAP,DTO6,DTO2,TDROP,UO,VO,SC,PRANTL,NSTOR,MIXED   6
     3  ,ASIDE(12),K1,E,E2,BETA,TSKIP,QBASE,FBASE,M1,M2,BTA,         MIXED   7
     4 BTD,BHS,BW,IMET,BLOW,F1,Q1,TD,TA,HS,W,G(1400,6),HEAT(20),     ITER   25
     5 FLOW(20),TH(20),NMET,NH,A,DTMET,TW,PR,DTDROP                  MIXED   9
     6 ,ASIDEH,HT,WID,ALEN,PB,ISPRAY                                 MIXED  10
      COMMON/SPSW/ TSPRON                                            MIXED  11
C     LOG-LINEAR INTERPOLATION OF HEAT TABLE                         MIXED  12
      DO 1 M1=M2,NH                                                  MIXED  13
      X1=X-TSKIP                                                     MIXED  14
      IF(X1.LE.0.0) GOTO 300                                         MIXED  15
      X9=ALOG(X1)                                                    MIXED  16
      IF(X9.LT.TH(M1)) GOTO 1                                        MIXED  17
      IF(X9.LT.TH(M1+1)) GOTO 1210                                   MIXED  18
    1 CONTINUE                                                       MIXED  19
 1210 F4=(X9-TH(M1))/(TH(M1+1)-TH(M1))                              MIXED  20
      M2=M1                                                          MIXED  21
C     EXTERNAL HEAT INPUT TO POND                                    MIXED  22
      Q1=HEAT(M1)+F4*(HEAT(M1+1)-HEAT(M1))                          MIXED  23
C     CIRCULATION THROUGH POND                                       MIXED  24
      F1=FLOW(M1)+F4*(FLOW(M1+1)-FLOW(M1))                          MIXED  25
C     ADD BASE HEAT LOAD AND FLOW, IF ANY                            MIXED  26
      Q1=Q1+QBASE                                                    MIXED  27
      F1=F1+FBASE                                                    MIXED  28
  300 CONTINUE                                                       MIXED  29
C     LINEAR INTERPOLATION OF MET TABLE                              MIXED  30
      IF(IMET.NE.0) GOTO 100                                         MIXED  31
      M1=X/DTMET+1                                                   MIXED  32
      F4=(X-(M1-1)*DTMET)/DTMET                                      MIXED  33
      TD=G(M1,1)+F4*(G(M1+1,1)-G(M1,1))                             MIXED  34
      TA=G(M1,2)+F4*(G(M1+1,2)-G(M1,2))                             MIXED  35
      HS=G(M1,3)+F4*(G(M1+1,3)-G(M1,3))                             MIXED  36
      W=G(M1,4)+F4*(G(M1+1,4)-G(M1,4))                              MIXED  37
      TW=G(M1,5)+F4*(G(M1+1,5)-G(M1,5))                             MIXED  38
      PB=G(M1,6)+F4*(G(M1+1,6)-G(M1,6))                             MIXED  39
      DATA WMIN/0.1/                                                 MIXED  40
C     MINIMUM WIND SPEED FOR CONTINUITY OF PROGRAM                   MIXED  41
      IF(W.LT.WMIN) W=WMIN                                           MIXED  42
  100 CONTINUE                                                       MIXED  43
      ETA=0.0                                                        MIXED  44
      FDR=0.0                                                        MIXED  45
      FEVAP=0.0                                                      MIXED  46
      HR=0.0                                                         MIXED  47
      HE=0.0                                                         MIXED  48
C     CALCULATE HEAT TRANSFER FROM SURFACE OF POND                   MIXED  49
      CALL EQTEMP(T,HR,HE)                                           MIXED  50
      IF(F1.LE.0.0)F1=1.0                                            JULY3046
      TSPRAY=T+Q1/(62.4*F1)                                         MIXED  51
```

Figure B.4 (Continued)

```
C        DELAY SPRAYS BY TSPRON HOURS                             MIXED 52
         IF(X.LT.TSPRON) GOTO 201                                 MIXED 53
         IF(ISPRAY.EQ.1) GOTO 200                                 MIXED 54
C     RIGOROUS MODEL                                              MIXED 55
         CALL SPRAY2(TSPRAY,ETA,FEVAP,FDR)                        MIXED 56
         GOTO 201                                                 MIXED 57
C     REGRESSION MODEL                                            MIXED 58
  200 CALL SPRAY(TSPRAY,ETA,FEVAP,FDR)                            MIXED 59
  201 CONTINUE                                                    MIXED 60
         HSPRAY=Q1-F1*62.4*ETA*(TSPRAY-TW)                        MIXED 61
C     RATE OF TEMPERATURE CHANGE, DEG F/HR                        MIXED 62
         FA=(HR*A/24+HSPRAY)/(62.4*V)                             MIXED 63
C     EVAPORATION RATE FROM SURFACE IN FT**3/HR                   MIXED 64
         DATA HVAP/1040.0/                                        MIXED 65
         E2=HE*A/(24*HVAP*62.4)                                   MIXED 66
         E2=E2+F1*(FEVAP+FDR)                                     MIXED 67
C     RATE OF VOLUME CHANGE, FT**3/HR                             MIXED 68
         FB=-BLOW-E2                                              MIXED 69
         RETURN                                                   MIXED 70
         END                                                      MIXED 71
         SUBROUTINE EQTEMP(T,HR,HE)                               EQTEMP  2
C     CALCULATE SURFACE HEAT TRANSFER AND EVAPORATION USING       EQTEMP  3
C     FORMULAE OF RYAN ET AL 1973                                 EQTEMP  4
         COMMON CH(6),CL(7),CEH(6),CEL(7),NORIFT,WDRO,DWDR,FDRIFT(20), EQTEMP  5
     1 CEMIN,CEMAX,CMIN,CMAX,VOL,AM,CON1,CON2,CON3,CON4,CON5,CON6, EQTEMP  6
     2 VIS,RHOA,DIFF,AK,H,EVAP,DTO6,OTO2,TDROP,UO,VO,SC,PRANTL,NSTDR, EQTEMP  7
     3 ATOP(12), ASIDE(12),K1,E,E2,BETA,TSKIP,QBASE,FBASE,M1,M2,BTA, EQTEMP  8
     4 BTD,BHS,BW,IMET,BLOW,F1,Q1,TD,TA,HS,W,G(1400,6),HEAT(20),   ITER  26
     5 FLOW(20),TH(20),NMET,NH,A,DTMET,TW,PR,DTDROP                EQTEMP10
     6 ,ASIDEH,HT,WID,ALEN,PB,ISPRAY                               EQTEMP11
         PAIR=PB*25.4                                             EQTEMP13
         DTV=(T+460)/(1-.378*PWAT(T)/PAIR)                        EQTEMP14
     1  -(TA+460)/(1-.378*PWAT(TD)/PAIR)                          EQTEMP15
         DTV3=0                                                   EQTEMP16
         IF(DTV.LE.0.0) GOTO 1500                                 EQTEMP17
         DTV3=DTV**.33333333                                      EQTEMP18
 1500 FU=(22.4*DTV3+14*W)                                         EQTEMP19
         HE=(PWAT(T)-PWAT(TD))*FU                                 EQTEMP20
         HC=C1*(T-TA)*FU                                          EQTEMP21
         DATA C1/0.26/                                            EQTEMP22
         HBR=4.026E-8*(460+T)**4                                  EQTEMP23
         HAN=1.16E-13*(TA+460)**6*(1-CC**2*.17)                   EQTEMP24
         DATA CC/0.0/                                             EQTEMP25
         HR=HS-HC+HAN-HBR-HE                                      EQTEMP26
         RETURN                                                   EQTEMP27
         END                                                      EQTEMP28
         FUNCTION PWAT(T)                                         PWAT   2
C     VAPOR PRESSURE OF AIR IN MM HG                              PWAT   3
C     FOR T IN DEG F                                              PWAT   4
     , TK=(T-32)/1.8+273.1                                        PWAT   5
         PWAT=760*EXP(71.02499-7381.6677/TK-9.0993037*ALOG(TK)    PWAT   6
     1 +.0070831558*TK)                                           PWAT   7
         RETURN                                                   PWAT   8
         END                                                      PWAT   9
         SUBROUTINE SPRAY(TSPRAY,ETA,FEVAP,FDR)                   SPRAY  2
C     SPRAY POND PERFORMANCE USING REGRESSION EQUATIONS           SPRAY  3
         COMMON CH(6),CL(7),CEH(6),CEL(7),NORIFT,WDRO,DWDR,FDRIFT(20), SPRAY  4
     1 CEMIN,CEMAX,CMIN,CMAX,VOL,AM,CON1,CON2,CON3,CON4,CON5,CON6, SPRAY  5
     2 VIS,RHOA,DIFF,AK,H,EVAP,DTO6,OTO2,TDROP,UO,VO,SC,PRANTL,NSTDR, SPRAY  6
     3 ATOP(12), ASIDE(12),K1,E,E2,BETA,TSKIP,QBASE,FBASE,M1,M2,BTA, SPRAY  7
     4 BTD,BHS,BW,IMET,BLOW,F1,Q1,TD,TA,HS,W,G(1400,6),HEAT(20),   ITER  27
     5 FLOW(20),TH(20),NMET,NH,A,DTMET,TW,PR,DTDROP                SPRAY  9
     6 ,ASIDEH,HT,WID,ALEN,PR,ISPRAY                               SPRAY 10
```

Figure B.4   (Continued)

```
      EQUIVALENCE(DB,TA)                                          SPRAY  11
C     HIGH WIND SPEED EFFICIENCY                                  SPRAY  12
      ETA=CH(1)+CH(2)*DB+CH(3)*TW+CH(4)*TSPRAY+CH(5)*W+CH(6)*SQRT(W)  SPRAY  13
C     LWS EFFICIENCY                                              SPRAY  14
      EL=CL(1)+CL(2)*DB+CL(3)*DB**2+CL(4)*DB**3+CL(5)*TW+          SPRAY  15
     1 CL(6)*TSPRAY+CL(7)*TSPRAY**2                               SPRAY  16
      IF(ETA.LT.EL) GOTO3                                         SPRAY  17
C     HIGH WIND SPEED EVAPORATION                                 SPRAY  18
      FEVAP=CEH(1)+CEH(2)*DB+CEH(3)*TW+CEH(4)*TSPRAY+             SPRAY  19
     1 CEH(5)*W+CEH(6)*SQRT(W)                                    SPRAY  20
      GOTO 4                                                      SPRAY  21
C     LOW WIND SPEED EVAPORATION                                  SPRAY  22
    3 FEVAP=CEL(1)+CEL(2)*DB+CEL(3)*DB**2+CEL(4)*DB**3+CEL(5)*TW   SPRAY  23
     1 +CEL(6)*TSPRAY+CEL(7)*TSPRAY**2                            SPRAY  24
      ETA=EL                                                      SPRAY  25
C     DRIFT LOSS                                                  SPRAY  26
    4 NTBL=(W-WDRO)/DWDR+1                                        SPRAY  27
      IF(NTBL.GE.NDRIFT) NTBL=NDRIFT-1                            SPRAY  28
      FDR=FDRIFT(NTBL)+((W-WDRO-(NTBL-1)*DWDR)/DWDR)*             SPRAY  29
     1 (FDRIFT(NTBL+1)-FDRIFT(NTBL))                              SPRAY  30
C     SET LIMITS ON EVAPORATION AND EFFICIENCY                    SPRAY  31
      IF(FEVAP.LT.CEMIN) FEVAP=CEMIN                              SPRAY  32
      IF(FEVAP.GT.CEMAX) FEVAP=CEMAX                              SPRAY  33
      IF(ETA.LT.CMIN) ETA=CMIN                                    SPRAY  34
      IF(ETA.GT.CMAX) ETA=CMAX                                    SPRAY  35
      RETURN                                                      SPRAY  36
      END                                                         SPRAY  37
      SUBROUTINE SPRAY2(THOT,ETA,FEVAP,FDR)                       SPRAY2  2
C     RIGOROUS SPRAY POND MODEL                                   SPRAY2  3
      DIMENSION TSEG(11),HUM(11)                                  SPRAY2  4
      COMMON CH(6),CL(7),CEH(6),CEL(7),NDRIFT,WDRO,DWDR,FDRIFT(20), SPRAY2  5
     1 CEMIN,CEMAX,CMIN,CMAX,VOL,AM,CON1,CON2,CON3,CON4,CON5,CON6, SPRAY2  6
     2 VIS,RHOA,DIFF,AK,H,EVAP,DTO6,DTO2,TDROP,UO,VO,SC,PRANTL,NSTOR, SPRAY2  7
     3 ATOP(12), ASIDE(12),K1,E,E2,BETA,TSKIP,QBASE,FBASE,M1,M2,BTA, SPRAY2  8
     4 BTD,BHS,BW,IMET,BLOW,F1,Q1,TD,TA,HS,W,G(1400,6),HEAT(20),   ITER   28
     5 FLOW(20),TH(20),NMET,NH,A,DTMET,TW,PR,DTDROP                SPRAY210
     6 ,ASIDEH,HT,WID,ALEN,PB,ISPRAY                              SPRAY211
      COMMON/DRPSZ/ R                                             SPRAY212
      EQUIVALENCE(TA,TDRY),(TW,TWET)                              SPRAY213
      C(Z)=(Z-32.)/1.8                                            SPRAY214
C     ALPHA IS CONVERGENCE PARAMETER OF LWS MODEL                 SPRAY215
      DATA ALPHA/-0.05/                                          AUG12   1
C  CONVERT MPH TO CM/SEC                                          SPRAY217
      WIND1=W*44.7                                                SPRAY218
C     CONVERT FLOW TO CC/SEC                                      SPRAY219
      Q=F1*7.87                                                   SPRAY220
C     DRIFT LOSS                                                  SPRAY221
    4 NTBL=(W-WDRO)/DWDR+1                                        SPRAY222
      IF(NTBL.GE.NDRIFT) NTBL=NDRIFT-1                            SPRAY223
      FDR=FDRIFT(NTBL)+((W-WDRO-(NTBL-1)*DWDR)/DWDR)*             SPRAY224
     1 (FDRIFT(NTBL+1)-FDRIFT(NTBL))                              SPRAY225
C     CALCULATE HUMIDITY                                          SPRAY226
      CALL PSY1(TDRY,TWET,PB,DP,PV,HUMID,ENTHAL,VOLUME,RH)        SPRAY227
      THOT1=C(THOT)                                               SPRAY228
      TDRY1=C(TDRY)                                               SPRAY229
      TWET1=C(TWET)                                               SPRAY230
C     HIGH WIND SPEED MODEL                                       SPRAY231
C     FOR LOW WIND SPEEDS, GOTO LWS MODEL DIRECTLY                SPRAY232
      IF(W.LT.3.0) GOTO2000                                       SPRAY233
      CALL HWS(THOT1,HUMID,TDRY1,         TWAV,WIND1,Q,R,EVAPS)    SPRAY234
C     HWS EFFICIENCY AND EVAPORATION                              SPRAY235
      ETA=(THOT1-TWAV)/(THOT1-TWET1)                              SPRAY236
      FEVAP=EVAPS/Q                                               SPRAY237
```

Figure B.4   (Continued)

```
 2000 CONTINUE                                                           SPRAY238
C         SKIP LWS MODEL FOR THIS CONDITION TO AVOID COMPUTATIONAL PROBLEMS  SPRAY239
C         HWS EFFICIENCY                                                 SPRAY240
          IF(TDRY.GT.THOT) GOTO 1111                                     SPRAY241
          DATA KOUNT/0/                                                  SPRAY242
          IF(KOUNT.GT.1) GOTO 445                                        SPRAY243
C         INITIALIZE HUMIDITY AND TEMPERATURE IF FIRST RUN               SPRAY244
          DO 444 L=2,11                                                  SPRAY245
          TSEG(L)=TDRY1+1.0                                              SPRAY246
  444     HUM(L)=HUMID+.01                                               SPRAY247
          KOUNT=KOUNT+1                                                  SPRAY248
  445     CONTINUE                                                       SPRAY249
C         LOW WIND SPEED MODEL                                           SPRAY250
          CALL LWS(THOT1,HUMID,TDRY1,TWAV,Q,R,TSEG,HUM,ALPHA,EVAPS)      SPRAY251
C         LWS EFFICIENCY AND EVAPORATION                                 SPRAY252
          ETA2=(THOT1-TWAV)/(THOT1-TWET1)                                SPRAY253
          FEVAP2=EVAPS/Q                                                 SPRAY254
C         PICK LARGER EFFICIENCY                                         SPRAY255
          IF(ETA.GT.ETA2) GOTO 1002                                     AUG12   2
          ETA=ETA2                                                       SPRAY257
          FEVAP=FEVAP2                                                   SPRAY258
C         LIMITS ON EFFICIENCY AND EVAPORATION                           SPRAY259
 1002     IF(ETA.GT.CMAX) ETA=CMAX                                       SPRAY260
          IF(ETA.LT.CMIN) ETA=CMIN                                       SPRAY261
          IF(FEVAP.LT.CEMIN) FEVAP=CEMIN                                 SPRAY262
          IF(FEVAP.GT.CEMAX) FEVAP=CEMAX                                 SPRAY263
          RETURN                                                         SPRAY264
C         FALL BACK ON REGRESSION MODEL                                  SPRAY265
 1111     CONTINUE                                                       SPRAY266
          CALL SPRAY(THOT,ETA,FEVAP,FDR)                                 SPRAY267
          RETURN                                                         SPRAY268
          END                                                           SPRAY269
          SUBROUTINE LWS(THOT,HUMID,TAIR,TWAV,Q,R,TSEG,HUM,ALPHA,EVAPS)  LWS    2
C         LOW WIND SPEED MODEL                                           LWS    3
          DIMENSION VUP(12),FLOW(12),QT(12),RHO2(12),VH(12)              LWS    4
          DIMENSION TSEG(11),HUM(11),HOUT(11)                            LWS    5
          DIMENSION HFIL(12),TFIL(12)                                    LWS    6
          DIMENSION TM2(12),TM1(12),HM2(12),HM1(12)                      LWS    7
          COMMON CH(6),CL(7),CEH(6),CEL(7),NDRIFT,WDRO,DWOR,FDRIFT(20),  LWS    8
         1 CEMIN,CEMAX,CMIN,CMAX,VOL,AM,CON1,CON2,CON3,CON4,CON5,CON6,   LWS    9
         2 VIS,RHOA,DIFF,AK,H,EVAP,DTO6,DTO2,TDROP,UO,VO,SC,PRANTL,NSTDR,LWS   10
         3 ATOP(12), ASIDE(12),K1,E,E2,BETA,TSKIP,QBASE,FBASE,M1,M2,BTA, LWS   11
         4 BTD,BHS,BW,IMET,BLOW,F1,Q1,TD,TA,HS,W,G(1400,6),HEAT(20),     ITER  29
         5 DUM1(20),TH(20),NMET,NH,A,DTMET,TW,PR,DTDROP                  LWS   13
         6 ,ASIDEH,HT,WID,ALEN,PB,ISPRAY                                 LWS   14
          DO 491 I=1,12                                                  LWS   15
          TM2(I)=0                                                       LWS   16
          TM1(I)=0                                                       LWS   17
          HM2(I)=0                                                       LWS   18
  491     HM1(I)=0                                                       LWS   19
          TLAST=0                                                        LWS   20
          DATA HVAP,CP,RHO/580.0,1.0,1.0/                                LWS   21
          ICNT=0                                                         LWS   22
C         DENSITY OF AMBIENT AIR GM/CC                                   LWS   23
          RHO1=(1+HUMID)/((81.86*TAIR+22387)*(.03448+HUMID/18))          LWS   24
          FLOW(11)=0                                                     LWS   25
          QT(1)=0                                                        LWS   26
          FLOW(1)=0                                                      LWS   27
          RHO2(1)=RHO1                                                   LWS   28
          ATOT=ALEN*WID                                                  LWS   29
          TSEG(1)=TAIR                                                   LWS   30
          HUM(1)=HUMID                                                   LWS   31
C         CONCENTRATION OF WATER IN AIR                                  LWS   32
```

Figure B.4   (Continued)

```
      CWA=HUMID/((81.86*TAIR+22387)*(.03448+HUMID/18))          LWS   33
C     BEGIN ITERATIVE SOLUTION                                   LWS   34
      DO 801 NITER=1,20                                          LWS   35
      DO 101 J=1,10                                              LWS   36
      I=12-J                                                     LWS   37
C     DENSITY OF AIR IN EACH SEGMENT GM/CC                       LWS   38
      RHO2(I)=(1+HUM(I))/((81.86*TSEG(I)+22387)*(.03448+HUM(I)/18)) LWS 39
C     HUMID VOLUME, CC/GM BDA                                    LWS   40
      VH(I)=((81.86*TSEG(I)+22387)*(.03448+HUM(I)/18))           LWS   41
  101 CONTINUE                                                   LWS   42
  105 CONTINUE                                                   LWS   43
      DO 1001 J=1,10                                             LWS   44
      I=12-J                                                     LWS   45
      DRHO=RHO1-RHO2(I)                                          LWS   46
      ARG=980*DRHO*HT*.5/RHO1                                    LWS   47
      ICNT=1                                                     LWS   48
      IF(ARG.LT.0.0) GOTO 668                                    LWS   49
C     UPWARD VELOCITY OF AIR LEAVING EACH SEGMENT                LWS   50
      VUP(I)=SQRT(ARG)                                           LWS   51
  668 CONTINUE                                                   LWS   52
C     MATERIAL BALANCE ON EACH SEGMENT                           LWS   53
      QT(I)=VUP(I)*ATOP(I)/VH(I)                                 LWS   54
      FLOW(I-1)=FLOW(I)+QT(I)                                    LWS   55
 1001 CONTINUE                                                   LWS   56
      ICNT=ICNT+1                                                LWS   57
  104 CONTINUE                                                   LWS   58
C     ENTHALPY OF AIR ENTERING FIRST SEGMENT, CAL/GM BDA         LWS   59
      HOUT(1)=FLOW(1)*(.238*TAIR+HUMID*(HVAP+.45*TAIR))          LWS   60
      TSEG(1)=TAIR                                               LWS   61
      EVAPS=0                                                    LWS   62
      HUM(1)=HUMID                                               LWS   63
      SUMTC=0                                                    LWS   64
      DO 201 I=2,11                                              LWS   65
      TEMP=TSEG(I-1)+273.2                                       LWS   66
C     VISCOSITY OF AIR, GM/(SEC CM)                              LWS   67
      VIS=2.7936E-6*TEMP**.73617                                 LWS   68
C     DENSITY OF AIR, GM/CC                                      LWS   69
      RHOA=.353/TEMP                                             LWS   70
C     DIFFUSION COEFF OF AIR(CM**2/SEC)                          LWS   71
      DIFF=5.8758E-6*TEMP**1.8615                                LWS   72
C     PRANTL NO                                                  LWS   73
      PRANTL=.93176*TEMP**(-.042784)                             LWS   74
C     SCHMIDT NO                                                 LWS   75
      SC=2.2705*TEMP**(-.21398)                                  LWS   76
C     THERMAL CONDUCTIVITY OF AIR,CM/SEC                         LWS   77
      AC=3.9273E-7*TEMP**.88315                                  LWS   78
      CON4=AC/R                                                  LWS   79
      CON6=2*R*RHOA/VIS                                          LWS   80
      CON5=DIFF/R                                                LWS   81
      TDROP=THOT                                                 LWS   82
C     CALCULATE TEMPERATURE AND EVAPORATION OF FALLING DROPS     LWS   83
      CALL DROP(TSEG(I-1),CWA)                                   LWS   84
C     SENSIBLE HEAT TRANSFER IN SEGMENT                          LWS   85
      HSEG=RHO*CP*(Q*ATOP(I)/ATOT)*(THOT-TDROP)                  LWS   86
C     EVAPORATION IN SEGMENT                                     LWS   87
      EVAP1=EVAP*Q*ATOP(I)/(ATOT*VOL)                            LWS   88
C     SENSIBLE HE AT LEAVING SEGMENT AND ENTERING NEXT           LWS   89
      HOUT(I)=HSEG+HOUT(I-1)*(1-QT(I-1)/(QT(I-1)+FLOW(I-1)))     LWS   90
C     HUMIDITY IN SEGMENT                                        LWS   91
      HUM(I)=HUM(I-1)+EVAP1/FLOW(I-1)                            LWS   92
C     TEMPERATURE IN SEGMENT                                     LWS   93
      TSEG(I)=(HOUT(I)/FLOW(I-1)-HUM(I)*HVAP)/(.238+.45*HUM(I))  LWS   94
      EVAPS=EVAPS+EVAP1                                          LWS   95
```

Figure B.4   (Continued)

```
          CWA=HUM(I)/((81.86*TSEG(I)+22387)*(.03448+HUM(I)/18))     LWS   96
          SUMTC=SUMTC+TDROP*ATOP(I)                                 LWS   97
  201 CONTINUE                                                      LWS   98
C         AVERAGE TEMPERATURE OF WATER FALLING TO POND SURFACE      LWS   99
          TWAV=SUMTC/ATOT                                           LWS  100
          IF(NITER.LT.3) GOTO 49                                    LWS  101
          DO 492 I=2,11                                             LWS  102
C         SECOND ORDER SMOOTHING OPERATOR TO AID CONVERGENCE        LWS  103
          HFIL(I)=ALPHA*(HM2(I)-2*HM1(I)+HUM(I))                    LWS  104
          TFIL(I)=ALPHA*(TM2(I)-2*TM1(I)+TSEG(I))                   LWS  105
  492 CONTINUE                                                      LWS  106
          DO 493 I=2,11                                             LWS  107
          TSEG(I)=TSEG(I)+TFIL(I)                                   LWS  108
          HUM(I)=HUM(I)+HFIL(I)                                     LWS  109
  493 CONTINUE                                                      LWS  110
   49 DO 494 I=2,11                                                 LWS  111
          TM2(I)=TM1(I)                                             LWS  112
          TM1(I)=TSEG(I)                                            LWS  113
          HM2(I)=HM1(I)                                             LWS  114
  494 HM1(I)=HUM(I)                                                 LWS  115
          IF(ABS((TLAST-TWAV)/TWAV).LT.0.002) GOTO 800             LWS  116
          TLAST=TWAV                                                LWS  117
  801 CONTINUE                                                      LWS  118
          WRITE(6,20)                                               LWS  119
   20 FORMAT(10X,'NO CONVERGENCE AFTER 20 TRIES')                   LWS  120
  800 RETURN                                                        LWS  121
          END                                                       LWS  122
          SUBROUTINE HWS(THOT,HUMID,TAIR,TWAV,WIND,Q,R,EVAPS)       HWS    2
C         HIGH WIND SPEED MODEL                                     HWS    3
          COMMON CH(6),CL(7),CEH(6),CEL(7),NDRIFT,WDRQ,DWDR,FDRIFT(20), HWS 4
     1 CEMIN,CEMAX,CMIN,CMAX,VOL,AM,CON1,CON2,CON3,CON4,CON5,CON6,  HWS    5
     2 VIS,RHOA,DIFF,AK,H,EVAP,DTO6,DTO2,TDROP,U0,V0,SC,PRANTL,NSTOR, HWS  6
     3 ATOP(12),ASIDE(12),K1,E,E2,BETA,TSKIP,QBASE,FBASE,M1,M2,BTA, HWS    7
     4 BTD,BHS,BW,IMET,BLOW,F1,G1,TD,TA,HS,W,G(1400,6),HEAT(20),    ITER  30
     5 FLOW(20),TH(20),NMET,NH,A,DTMET,TW,PR,DTDROP                 HWS    9
     6 ,ASIDEH,HT,WID,ALEN,PB,ISPRAY                                HWS   10
          DIMENSION TSEG(11),HUM(11),HOUT(11)                       HWS   11
          DATA HVAP,CP,RHO/580.0,1.0,1.0/                           HWS   12
          CON7=RHO*CP*Q/10                                          HWS   13
          CON8=Q/(10*VOL)                                           HWS   14
C         GMS OF BDA ENTERING SPRAY FIELD FROM UPWIND               HWS   15
          FLO=WIND*ASIDEH/((81.86*TAIR+22387)*(.03448+HUMID/18))    HWS   16
C         ENTHALPY OF AIR ENTERING SPRAY FIELD,CAL/SEC              HWS   17
          HOUT(1)=FLO *(.238*TAIR+HUMID*(HVAP+.45*TAIR))            HWS   18
          TSEG(1)=TAIR                                              HWS   19
          HUM(1)=HUMID                                              HWS   20
C         CONCENTRATION OF WATER IN AIR                             HWS   21
          CWA=HUMID/((81.86*TAIR+22387)*(.03448+HUMID/18))          HWS   22
          EVAPS=0                                                   HWS   23
          SUMTC=0                                                   HWS   24
          DO 1 I=2,11                                               HWS   25
          TEMP=TSEG(I-1)+273.2                                      HWS   26
C         VISCOSITY OF AIR GM/(CM SEC)                              HWS   27
          VIS=2.7936E-6*TEMP**.73617                                HWS   28
C         DENSITY OF AIR GM/CC                                      HWS   29
          RHOA=.353/TEMP                                            HWS   30
C         DIFFUSION COEFFICIENT OF AIR CM**2/SEC                    HWS   31
          DIFF=5.8758E-6*TEMP**1.8615                               HWS   32
C         PRANTL NO                                                 HWS   33
          PRANTL=.93176*TEMP**(-.042784)                            HWS   34
C         SCHMIDT NO                                                HWS   35
          SC=2.2705*TEMP**(-.21398)                                 HWS   36
C         THERMAL CONDUCTIVITY OF AIR CM/SEC                        HWS   37
```

Figure B.4   (Continued)

```
          AC=3.9273E-7*TEMP**.88315                                    HWS    38
          CON4=AC/R                                                    HWS    39
          CON6=SQRT(2*R*RHOA/VIS)                                      HWS    40
          CON5=DIFF/R                                                  HWS    41
          TDROP=THOT                                                   HWS    42
C         TEMPERATURE AND EVAPORATION OF DROP                          HWS    43
          CALL DROP(TSEG(I-1),CWA)                                     HWS    44
C         SENSIBLE HEAT ENTERING SEGMENT FROM DROPS                    HWS    45
          HSEG=CON7*(THOT-TDROP)                                       HWS    46
C         EVAPORATION FROM ALL DROPS INTO SEGMENT                      HWS    47
          EVAP1=EVAP*CON8                                              HWS    48
C         ENTHALPY LEAVING SEGMENT AND ENTERING NEXT                   HWS    49
          HOUT(I)=HOUT(I-1)+HSEG                                       HWS    50
C         HUMIDITY OF SEGMENT                                          HWS    51
          HUM(I)=HUM(I-1)+EVAP1/FLO                                    HWS    52
C         AIR TEMPERATURE IN SEGMENT                                   HWS    53
          TSEG(I)=(HOUT(I)/FLO -HUM(I)*HVAP)/(.24+.45*HUM(I))          HWS    54
          EVAPS=EVAPS+EVAP1                                            HWS    55
C    CWA = CONCENTRATION OF WATER IN AIR, GM/CC                        HWS    56
          CWA=HUM(I)/((81.86*TSEG(I)+22387)*(.03448+HUM(I)/18))        HWS    57
          SUMTC=SUMTC+TDROP                                            HWS    58
        1 CONTINUE                                                     HWS    59
C         AVERAGE TEMPERATURE OF WATER FALLING TO POND SURFACE         HWS    60
          TWAV=SUMTC/10                                                HWS    61
          RETURN                                                       HWS    62
          END                                                          HWS    63
          SUBROUTINE DROP(TAIR,CINF)                                   DROP    2
          COMMON CH(6),CL(7),CEH(6),CEL(7),NDRIFT,WDRO,DWDR,FDRIFT(20), DROP   3
         1 CEMIN,CEMAX,CMIN,CMAX,VOL,AM,CON1,CON2,CON3,CON4,CON5,CON6,  DROP   4
         2 VIS,RHOA,DIFF,AK,H,EVAP,DTO6,DTO2,TDROP,UO,VO,SC,PRANTL,NSTDR,DROP   5
         3 ATOP(12), ASIDE(12),K1,E,E2,BETA,TSKIP,QBASE,FBASE,M1,M2,BTA, DROP  6
         4 BTD,BHS,BW,IMET,BLOW,F1,Q1,TD,TA,HS,W,G(1400,6),HEAT(20),     ITER  31
         5 FLOW(20),TH(20),NMET,NH,A,DTMET,TW,PR,DTDROP                  DROP   8
         6 ,ASIDEH,HT,WID,ALEN,PB,ISPRAY                                DROP    9
C         CALCULATE HEAT AND MASS TRANSFER FROM A DROP                 DROP   10
          EVAP=0                                                       DROP   11
          ICNT=1                                                       DROP   12
C            BEGIN FOURTH ORDER RUNGE-KUTTA INT.OF EQUATIONS           DROP   13
          DO 1 I=1,NSTDR                                               DROP   14
          CALL FTDROP(ICNT,TDROP,DTD1,DI1,TAIR,CINF)                   DROP   15
          ICNT=ICNT+1                                                  DROP   16
          TDROP1=TDROP+DTO2*DTD1                                       DROP   17
          CALL FTDROP(ICNT,TDROP1,DTD2,DI2,TAIR,CINF)                  DROP   18
          TDROP2=TDROP+DTO2*DTD2                                       DROP   19
          CALL FTDROP(ICNT,TDROP2,DTD3,DI3,TAIR,CINF)                  DROP   20
          ICNT=ICNT+1                                                  DROP   21
          TDROP3=TDROP+DTD3*DTDROP                                     DROP   22
          CALL FTDROP(ICNT,TDROP3,DTD4,DI4,TAIR,CINF)                  DROP   23
          TDROP=TDROP+(DTD1+2*(DTD2+DTD3)+DTD4)*DTO6                   DROP   24
          EVAP=EVAP+(DI1+2*(DI2+DI3)+DI4)*DTO6                         DROP   25
        1 CONTINUE                                                     DROP   26
          RETURN                                                       DROP   27
          END                                                          DROP   28
          SUBROUTINE FTDROP(ICNT,TDRP,DTO,DI,TAIR,CINF)                FTDROP  2
          COMMON CH(6),CL(7),CEH(6),CEL(7),NDRIFT,WDRO,DWDR,FDRIFT(20), FTDROP 3
         1 CEMIN,CEMAX,CMIN,CMAX,VOL,AM,CON1,CON2,CON3,CON4,CON5,CON6,  FTDROP 4
         2 VIS,RHOA,DIFF,AK,H,EVAP,DTO6,DTO2,TDROP,UO,VO,SC,PRANTL,NSTDR,FTDROP 5
         3 ATOP(12), ASIDE(12),K1,E,E2,BETA,TSKIP,QBASE,FBASE,M1,M2,BTA, FTDROP 6
         4 BTD,BHS,BW,IMET,BLOW,F1,Q1,TD,TA,HS,W,G(1400,6),HEAT(20),     ITER  32
         5 FLOW(20),TH(20),NMET,NH,A,DTMET,TW,PR,DTDROP                  FTDROP 8
         6 ,ASIDEH,HT,WID,ALEN,PB,ISPRAY                                FTDROP 9
C         RATE OF HEAT AND MASS TRANSFER FROM A DROP                   FTDROP10
          COMMON/RESTOR/ SQV(100)                                      FTDROP11
```

Figure B.4   (Continued)

```
      DATA RG/82.02/                                              FTDROP12
      TDK=TDRP+273.2                                              FTDROP13
C     VAPOR PRESSURE OF WATER ATM                                 FTDROP14
      P=EXP(71.02499-7361.6477/TDK-9.0993037*ALOG(TDK)            FTDROP15
     1 +.0070831558*TDK)                                          FTDROP16
      SRE=CON6*SQV(ICNT)                                          FTDROP17
      HC=CON4*(1+.3*PRANTL**.3333333*SRE)                         FTDROP18
      HD=CON5*(1+.3*SC**.3333333*SRE)                             FTDROP19
      CDROP=P*18.0/(RG*TDK)                                       FTDROP20
C     RATE OF MASS TRANSFER                                       FTDROP21
      DI=CON3*HD*(CDROP-CINF)                                     FTDROP22
      DATA HVAP/580.0/                                            FTDROP23
C     RATE OF TEMPERATURE CHANGE                                  FTDROP24
      DTD=-CON1*(DI*HVAP+CON3*HC*(TDRP-TAIR))                     FTDROP25
      RETURN                                                      FTDROP26
      END                                                         FTDROP27
      SUBROUTINE INIT(R,THETA,YO,VELO)                            INIT    2
C     INITIALIZE CONSTANTS AND VELOCITIES OF BALLISTIC DROP       INIT    3
      COMMON CH(6),CL(7),CEH(6),CEL(7),NDRIFT,WDRO,DWDR,FDRIFT(20),INIT   4
     1 CEMIN,CEMAX,CMIN,CMAX,VOL,AM,CON1,CON2,CON3,CON4,CON5,CON6, INIT   5
     2 VIS,RHOA,DIFF,AK,H,EVAP,DTO6,DTO2,TDROP,UO,VO,SC,PRANTL,NSTDR,INIT 6
     3 ATOP(12), ASIDE(12),K1,E,E2,BETA,TSKIP,QBASE,FBASE,M1,M2,BTA,INIT  7
     4 BTD,BHS,BW,IMET,BLOW,F1,Q1,TD,TA,MS,W,Z(8400  ),HEAT(20),   ITER   33
     5 FLOW(20),TH(20),NMET,NH,A,DTMET,TW,PR,DTDROP                INIT    9
     6 ,ASIDEH,HT,WID,ALEN,PB,ISPRAY                               INIT   10
      COMMON/RESTOR/SQV(100)                                      INIT   11
      VOL=(3.1415926*4/3)*R**3                                    INIT   12
      DATA G/980.0/                                               INIT   13
      DATA HVAP,CP,RHO/580.0,1.0,1.0/                             INIT   14
      A=3.1415926*R**2                                            INIT   15
      CON1=1.0/VOL                                                INIT   16
      CON2=HVAP*12.566371*R**2                                    INIT   17
      CON3=12.566371*R**2                                         INIT   18
      VO=VELO*SIN(THETA)                                          INIT   19
      UO=VELO*COS(THETA)                                          INIT   20
C     TIME FOR DROP TO HIT SURFACE OF WATER                       INIT   21
      TFALL=VO/G+SQRT((VO/G)**2+2*YO/G)                           INIT   22
      DTDROP=TFALL/NSTDR                                          INIT   23
      DTO6=DTDROP/6                                               INIT   24
      DTO2=DTDROP/2                                               INIT   25
      NUM=NSTDR*2+10                                              INIT   26
      DO 1 I=1,NUM                                                INIT   27
      T=(I-1)*DTO2                                                INIT   28
C     VELOCITY OF DROP                                            INIT   29
      V=SQRT(UO**2+(VO-980*T)**2)                                 INIT   30
    1 SQV(I)=SQRT(V)                                              INIT   31
      RETURN                                                      INIT   32
      END                                                         INIT   33
      SUBROUTINE PSY1(DB,WB,PB,DP,PV,W,H,V,RH)                    PSY1    2
C     THIS ROUTINE CALCULATES' VAPOR PRESSURE PV, HUMIDITY RATIO W,PSY1   3
C         ENTHALPY H, VOLUME V, RELATIVE HUMIDITY RH, AND         PSY1    4
C         DEW POINT TEMPERATURE DP\                               PSY1    5
C         WHEN THE DRY BULB TEMPERATURE DB, WET BULB TEMPERATURE WB,PSY1  6
C         AND BAROMETRIC PRESSURE PB ARE GIVEN                    PSY1    7
C     UNITS' DB, WB, + DP )F>\ PB, + PV )IN OF HG>\ W)= WATER VAPOR PSY1  8
C         PER = DRY AIR>\ H )BTU/= OF DRY AIR>\ V )FT**3/= OF DRY PSY1    9
C         AIR\ RH IS A FRACTION, NOT (                            PSY1   10
      C(F)=(F-32.0E0)/1.8E0                                       PSY1   11
      PVP=PVSF(WB)                                                PSY1   12
      WSTAR=0.622*PVP/(PB-PVP)                                    PSY1   13
      IF (WB.GT.32.0) GO TO 105                                   PSY1   14
      PV=PVP-5.704E-4*PB*(DB-WB)/1.8                              PSY1   15
      GO TO 110                                                   PSY1   16
```

Figure B.4   (Continued)

```
105 CDB=C(DB)                                                    PSY1  17
    CWB=C(WB)                                                    PSY1  18
    HL=597.31+0.4409*CDB-CWB                                     PSY1  19
    CH=0.2402+0.4409*WSTAR                                       PSY1  20
    EX=(WSTAR-CH*(CDB-CWB)/HL)/0.622                             PSY1  21
    PV=PB*EX/(1.+EX)                                             PSY1  22
110 W=0.622*PV/(PB-PV)                                           PSY1  23
    V=0.754*(DB+459.7)*(1.0+7000.0*W/4360.0)/PB                 PSY1  24
    H=0.24*DB+(1061.0+0.444*DB)*W                               PSY1  25
    IF (PV.GT.0.0) GO TO 115                                     PSY1  26
    PV=0.0                                                       PSY1  27
    DP=0.0                                                       PSY1  28
    RH=0.0                                                       PSY1  29
    RETURN                                                       PSY1  30
115 IF (DB.NE.WB) GO TO 120                                      PSY1  31
    DP=DB                                                        PSY1  32
    RH=1.0                                                       PSY1  33
    RETURN                                                       PSY1  34
120 DP=DPF(PV)                                                   PSY1  35
    RH=PV/PVSF(DB)                                               PSY1  36
    RETURN                                                       PSY1  37
    END                                                          PSY1  38
    FUNCTION PVSF(X)                                             PSY1  39
    DIMENSION A(6),B(4),P(4)                                     PSY1  40
    DATA A/-7.90298,5.02808,-1.3816E-7,11.344,8.1328E-3,-3.49149/ PSY1 41
    DATA B/-9.09718,-3.56654,0.876793,0.0060273/               PSY1  42
    T=(X+459.688)/1.8                                            PSY1  43
    IF (T.LT.273.16) GO TO 100                                   PSY1  44
    Z=373.16/T                                                   PSY1  45
    P(1)=A(1)*(Z-1.0)                                            PSY1  46
    P(2)=A(2)*ALOG10(Z)                                          PSY1  47
    Z1=A(4)*(1.0-1.0/Z)                                          PSY1  48
    P(3)=A(3)*(10.0**Z1-1.0)                                     PSY1  49
    Z1=A(6)*(Z-1.0)                                              PSY1  50
    P(4)=A(5)*(10.0**Z1-1.0)                                     PSY1  51
    GO TO 105                                                    PSY1  52
100 Z=273.16/T                                                   PSY1  53
    P(1)=B(1)*(Z-1.0)                                            PSY1  54
    P(2)=B(2)*ALOG10(Z)                                          PSY1  55
    P(3)=B(3)*(1.0-1.0/Z)                                        PSY1  56
    P(4)=ALOG10(B(4))                                            PSY1  57
105 SUM=0.0                                                      PSY1  58
    DO 110 I=1,4                                                 PSY1  59
110 SUM=SUM+P(I)                                                 PSY1  60
    PVSF=29.921*10.0**SUM                                        PSY1  61
    RETURN                                                       PSY1  62
    END                                                          PSY1  63
    FUNCTION DPF(PV)                                             PSY1  64
C   THIS ROUTINE CALCULATES DEW-POINT TEMPERATURE FOR A GIVEN    PSY1  65
C        VAPOR PRESSURE PV                                       PSY1  66
    DP(A,B,C,Y)=A+(B+C*Y)*Y                                      PSY1  67
    Y=ALOG(PV)                                                   PSY1  68
    IF (PV.GT.0.1836) GO TO 100                                  PSY1  69
    DPF=DP(71.98,24.873,0.8927,Y)                               PSY1  70
    RETURN                                                       PSY1  71
100 DPF=DP(79.047,30.579,1.8893,Y)                              PSY1  72
    RETURN                                                       PSY1  73
    END                                                          PSY1  74
```

Figure B.4   (Continued)

```
      PROGRAM COMET2(INPUT,OUTPUT,TAPE5=INPUT,TAPE6=OUTPUT)

C
C     SPRAY POND DATA COMPARISON MODEL
C     COMPARE WATER USAGE AND TEMPERATURE FOR TWO SETS OF METEOROLOGY
C     RICHARD CODELL - US NRC, WASHINGTON DC, DECEMBER 1979
C                                                                    000110
C     TW1= WET BULB TEMPERATURE FOR DATA SET 1
C     TA1= DRY BULB TEMP. FOR DATA SET 1      (F)                    000130
C     W1= WIND SPEED FOR DATA SET 1     (MPH)                        000140
C     H1= RATE OF INSOLATION FOR DATA SET 1    (BTU/FT**2/DAY)       000150
C     TW2= WET BULB TEMPERATURE FOR DATA SET 2
C     TA2= DRY BULB TEMP. FOR DATA SET 2     (F)                     000170
C     W2= WIND SPEED FOR DATA SET 2    (MPH)                         000180
C     H2= RATE OF INSOLATION FOR DATA SET 2     (BTU/FT**2/DAY)      000190
C     PB1 = BAROMETRIC PRESSURE, DATA SET 1(INCHES MERCURY)
C     PB2 = BAROMETRIC PRESSURE, DATA SET 2(INCHES MERCURY)
C                                                                    000200
      COMMON HE,  FEVAP,FDR,WDRO,NDRIFT,DWDR,FDRIFT(20),CH(6),CL(7),
     1 CEH(6),CEL(7),HEAT,CON1,CON2,CON3,DTSPRY,DTIME,QSPRAY,V,TD
      DATA QX,QY,QX2,QY2,QCROSS/5*0.0/                               000270
      DATA ERR/1.0E-30/                                             000280
      DATA SX,SY,SX2,SY2,SCROSS/5*0./                                000290
      NAMELIST /INLIST/ DTIME,V,A,QSPRAY,HEAT,NDRIFT,WDRO,DWDR,FDRIFT
      PRINT 95
   95 FORMAT(1H1,20X,'DIFFERENCES IN STEADY STATE TEMPERATURES AND WATER
     1 USE FOR SUBJECT SPRAY POND',/20X,'USING MONTHLY AVERAGE VALUES OF
     2 WET BULB,DRY BULB,WIND SPEED,AND SOLAR RADIATION FROM ONSITE
     3',/20X'AND OFFSITE MET STATIONS',///)
      DTIME=0.0
      HEAT=5.0E8
      NDRIFT=2
      WDRO=0
      DWDR=2
      FDRIFT(1)=.0000001
      FDRIFT(2)=.0000001
C     COEFFICIENTS FOR MULTIPLE REGRESSION MODELS OF SPRAY EFFICIENCY
C     AND EVAPORATION LOSS GENERATED BY PROGRAM SPRCO
      READ(5,555) CH,CL,CEH,CEL
  555 FORMAT(4E15.8)
      READ(5,INLIST)
C     ESTIMATE ITERATION TIME IF NOT SPECIFIED
      IF(DTIME.GT.0.0) GOTO 40
      DTIME=10.0*HEAT/(62.4*V)
   40 CONTINUE
      WRITE(6,50) DTIME,V,A,QSPRAY,HEAT,WDRO,DWDR
   50 FORMAT(//20X,'TIMESTEP IN ITERATION DTIME = ',F10.3,' HOURS'/
     1 20X,'VOLUME OF POND, V = ',F12.1,' FT**3'/
     1 20X,'SURFACE AREA OF POND, A = ',F12.1,' FT**2'/
     2 20X,'RATE OF SPRAYING, QSPRAY = ',F12.1,' FT**3/SEC'/
     3 20X,'STEADY HEAT LOAD, HEAT = ',F12.1,' BTU/HR'/
     5 20X,'LOWER LIMIT OF WIND IN DRIFT TABLE WDRO = ',F10.2,' MPH'/
     6 20X,'INCREMENT IN DRIFT TABLE,DWDR = ',F10.2,' MPH'//)
      WRITE(6,52)
   52 FORMAT(//,15X,'DRIFT LOSS TABLE',//,T18,'WIND SPEED, MPH',T34,'DRIF
     1T LOSS FRACTION',/)
      DO 51 I=1,NDRIFT
      WSP=(I-1)*DWDR+WDRO
   51 WRITE(6,53) WSP,FDRIFT(I)
   53 FORMAT(T20,F10.2,T40,F11.8)                                   000350
      DTSPRY=HEAT/(QSPRAY*3600*62.4)                                000360
      CON1=A/(1498*V)                                               000370
      CON2=A/(1497600)                                              000380
      CON3=62.4*3600
```

Figure B.5   Listing of program COMET2

```
      READ(5,499) I                                                  000390
  499 FORMAT(I2)                                                      000400
      DO 2 J=1,I                                                      000410
      READ(5,500) TW1,TA1,W1,H1,PB1,TW2,TA2,W2,H2,PB2
  500 FORMAT(10F8.0)
C
C        IF DATA ARE MISSING IN SECOND SET, SET EQUAL TO VALUE IN 1ST SET
C
      IF(TW2.EQ.0.0)TW2=TW1
      IF(TA2.EQ.0.0)TA2=TA1
      IF(W2.EQ.0.0)W2=W1
      IF(H2.EQ.0. ) H2=H1                                             000440
      IF(PB2.EQ.0.0)PB2=PB1
C                                                                     000450
C        CALCULATE STEADY STATE TEMPERATURE AND EVAPORATION RATE
C        FOR EACH DATA SET
C                                                                     000470
      E1=E(TA1,W1,H1,PB1,TW1)
      EVAP1=30*HE/(62.4*HVAP)
      EVAP1=EVAP1+30*(FDR+FEVAP)*QSPRAY*86400/A                       000500
      EVAP1=EVAP1*A
      E2=E(TA2,W2,H2,PB2,TW2)
      DATA HVAP/1040.0/
      EVAP2=30*HE/(62.4*HVAP)
      EVAP2=EVAP2+30*(FDR+FEVAP)*QSPRAY*86400/A                       000530
      EVAP2=EVAP2*A
      DE=E2-E1                                                        000540
      DEVAP=EVAP2-EVAP1                                               000550
      WRITE(6,99)
      WRITE(6,101) TW1,TA1,W1,H1,PB1,E1,EVAP1
      WRITE(6,200) TW2,TA2,W2,H2,PB2,E2,EVAP2
   99 FORMAT(T21,'WET BULB',T37,'DRY BULB',T51,'WIND SPEED',/T22,
     1 'SOLAR RAD.'T84,'PB',T97,'POND TEMP',T114,'EVAPORATION',/T22,
     2 '(DEG. F)',T36,'(DEG.F)',T54,'(MPH)',T80,'INCHES HG',T64,
     3 '(BTU/FT**2/DY)',T96,'(DEG. F)',T112,'     FT**3'//)
  101 FORMAT( 5X,'DATA SET 1',F12.2,5F15.2,F20.2,/)
  200 FORMAT( 5X,'DATA SET 2',F12.2,5F15.2,F20.2,/)
      WRITE(6,102) DE,DEVAP
  102 FORMAT(T77,'E2-E1 = ', F6.3,5X,'EVAP2-EVAP1 = ',F12.1)
C                                                                     000660
C        CALCULATE SUMS FOR CORRELATION COEFFICIENTS                  000670
C                                                                     000680
      SX=SX+E1                                                        000690
      SX2=SX2+E1**2                                                   000700
      SY=SY+E2                                                        000710
      SY2=SY2+E2**2                                                   000720
      SCROSS=SCROSS+E1*E2                                             000730
      QX=QX+EVAP1                                                     000740
      QX2=QX2+EVAP1**2                                                000750
      QY=QY+EVAP2                                                     000760
      QY2=QY2+EVAP2**2                                                000770
      QCROSS=QCROSS+EVAP1*EVAP2                                       000780
C                                                                     000790
C        DIFFERENCES IN EQUILIBRIUM TEMP DUE TO EACH PARAMETER.       000800
C                                                                     000810
      DTW=E(TA1,W1,H1,PB1,TW2)-E1
      DTA=E(TA2,W1,H1,PB1,TW1)-E1
      DW=E(TA1,W2,H1,PB1,TW1)-E1
      DH=E(TA1,W1,H2,PB1,TW1)-E1
      DPB=E(TA1,W1,H1,PB2,TW1)-E1
      DTOT=DTW+DTA+DW+DH+DPB
      WRITE(6,5)                                                      000870
    5 FORMAT(//10X, 'DIFFERENCES IN E BETWEEN DATA SET 2 AND DATA SET 1 000880
     1BY PARAMETER',/)                                                000890
```

Figure B.5   (Continued)

```
      WRITE(6,6)DTW
    6 FORMAT(10X,'DIFFERENCE DUE TO WET BULB = ',T50,F10.3,' DEG. F')     000920
      WRITE(6,7)DTA
    7 FORMAT(10X,'DIFFERENCE DUE TO DRY BULB TEMP. = ',T50,F10.3,' DEG.    000940
     1F')                                                                 000950
      WRITE(6,8) DW
    8 FORMAT(10X,'DIFFERENCE DUE TO WIND SPEED = ',T50,F10.3,' DEG. F')    000970
      WRITE(6,9)DH
    9 FORMAT(10X,'DIFFERENCE DUE TO INSOLATION = 'T50,F10.3,' DEG. F')
      WRITE(6,11) DPB
   11 FORMAT(10X,'DIFFERENCE DUE TO BAROMETRIC PRESSURE = ',T50,F10.3,
     1 ' DEG F')                                                          000990
      WRITE(6,10)DTOT
   10 FORMAT(10X,'SUMMATION OF INDIVIDUAL DIFFERENCES = ',T50,F10.3,' DE   001010
     1G. F',//,1X,130('*'),///)                                           001020
    2 CONTINUE                                                            001030
C                                                                        001040
C     CORRELATION ANALYSIS                                               001050
C                                                                        001060
      SXX=I*SX2-SX**2                                                     001070
      SYY=I*SY2-SY**2                                                     001080
      SXY=I*SCROSS-SX*SY                                                  001090
      RSQ=(SXY**2+ERR)/(SXX*SYY+ERR)                                     001100
      QXX=I*QX2-QX**2                                                     001110
      QYY=I*QY2-QY**2                                                     001120
      QXY=I*QCROSS-QX*QY                                                  001130
      QRSQ=(QXY**2+ERR)/(QXX*QYY+ERR)
      SERR=SQRT(((SXX*SYY)-SXY**2)/(I*(I-2)*SXX))
      QSERR=SQRT(((QXX*QYY)-QXY**2)/(I*(I-2)*QXX))
      WRITE(6,300) RSQ,SERR
      WRITE(6,310) QRSQ,QSERR
  300 FORMAT(10X,'SAMPLE R SQUARED FOR EQUILIBRIUM TEMP. = ',F10.3,
     1 10X,'STANDARD ERROR = ',F10.3,' DEG.F')
  310 FORMAT(10X,'SAMPLE R SQUARED FOR EVAPORATION = ',F10.3,
     1 10X,'STANDARD ERROR = ',F10.3,'FT**3')                            001180
      SXXI=SX /I                                                         001190
      SYYI=SY /I                                                         001200
      BIAS=SYYI-SXXI
      WRITE(6,250) SXXI,SYYI,BIAS
  250 FORMAT(10X,'AVERAGE E, DATA SET 1 = ',F12.3,/,10X,'AVERAGE E, DATA  001220
     1 SET 2 = ',F12.3,/,10X,'AVERAGE E2 - AVERAGE E1 = ',F12.4)         001230
      EBIAS=(QY-QX)/I                                                     001240
      WRITE(6,251) EBIAS
  251 FORMAT(10X, 'AVERAGE EVAP2 - AVERAGE EVAP1 = ',F12.4)              001260
      STOP                                                               001270
      END                                                                001280
      FUNCTION E(TA,W,H,PB,WB)                                           001300
C
C     CALCULATES THE STEADY STATE TEMPERATURE BY
C     AN ITERATIVE PROCESS, WITH SPRAY HEAT LOSS, EVAPORATION, AND
C     DRIFT DETERMINED BY REGRESSION COEFFICIENTS FROM PROGRAMS
C     #SPRAYCO# AND #DRIFT#                                              001330
C
      COMMON HE,  FEVAP,FDR,WDRO,NDRIFT,DWOR,FDRIFT(20),CH(6),CL(7),
     1 CEH(6),CEL(7),HEAT,CON1,CON2,CON3,DTSPRY,DTIME,QSPRAY,V,TD
      ES=100
C     CONVERT ATM PRESSURE TO MM
      PAIR=PB*760.0/29.92
C     CALCULATE DEW POINT TEMPERATURE
      CALL PSY1(TA,WB,PB,TD,PV,HUMRAT,ENTHAL,HUMVOL,RH)
C     BEGIN ITERATIVE SOLUTION FOR POND TEMPERATURE                      001430
      DO 1 I=1,50                                                        001440
      TSPRAY=ES+DTSPRY
C     SURFACE HEAT TRANSFER AND EVAPORATION FROM RYAN, 1973
```

Figure B.5   (Continued)

```
          DTV=(ES+460)/(1.-.378*PWAT(ES)/PAIR)-
        1 (TA+460)/(1.0-.378*PWAT(TD)/PAIR)
          DTV3=0
          IF(DTV.LE.0.0) GOTO 1500
          DTV3=DTV**.33333333
     1500 FU=22.4*DTV3+14*W
          HC=0.26*(ES-TA)*FU
          HBR=4.026E-8*(460+ES)**4
          HE=(PWAT(ES)-PWAT(TD))*FU
          HAN=1.16E-13*(TA+460)**6*(1.0-CC**2*.17)
C     CONSERVATIVE VALUE FOR CLOUD COVER
          DATA CC/0.0/
          HR=H-HC-HAN-HBR-HE
C     HWS EFFICIENCY
          ETA=CH(1)+CH(2)*TA+CH(3)*WB+CH(4)*TSPRAY+CH(5)*W+CH(6)*SQRT(W)
C     LWS EFFICIENCY
          EL=CL(1)+CL(2)*TA+CL(3)*TA**2+CL(4)*TA**3+CL(5)*WB+
        1 CL(6)*TSPRAY+CL(7)*TSPRAY**2
          IF(ETA.LT.EL) ETA=EL                                       001520
          IF(ETA.LT.0.0) ETA=0.0
          IF(ETA.GT.1.0) ETA=1.0
C     SPRAY HEAT LOSS
          HSPRAY=HEAT=QSPRAY*CON3*ETA*(TSPRAY-WB)                    001530
          DTEMP=HR*CON1+HSPRAY/(62.4*V)                             001550
          T1=ES                                                      001550
          ES=ES+DTEMP*DTIME                                          001560
          IF(ABS(T1-ES).LT.0.002) GO TO 2
        1 CONTINUE                                                   001580
        2 CONTINUE                                                   001590
          E=ES
          IF(ETA.EQ.EL) GOTO 3                                       001600
C     HIGH WIND SPEED EVAPORATION
          FEVAP=CEH(1)+CEH(2)*TA+CEH(3)*WB+CEH(4)*TSPRAY+            001620
        1 CEH(5)*W+CEH(6)*SQRT(W)
          GOTO 4                                                     001640
C     LOW WIND SPEED EVAPORATION
        3 FEVAP=CEL(1)+CEL(2)*TA+CEL(3)*TA**2+CEL(4)*TA**3+CEL(5)*WB
        1 +CEL(6)*TSPRAY+CEL(7)*TSPRAY**2
C     DRIFT LOSS
        4 NTBL=(W-WDR0)/DWDR+1                                       001670
          IF(NTBL.GE.NDRIFT) NTBL=NDRIFT-1                          001680
          FDR=FDRIFT(NTBL)+((W-WDR0-(NTBL-1)*DWDR)/DWDR)*           001690
        1 (FDRIFT(NTBL+1)-FDRIFT(NTBL))                            001700
          IF(FEVAP.LT.0.0) FEVAP=0.0
          IF(FEVAP.GT.1.0) FEVAP=1.0
          RETURN                                                     001710
          END                                                        001720
          FUNCTION PWAT(T)
          TK=(T-32.0)/1.8+273.1
          PWAT=760*EXP(71.02499-7381.6677/TK-9.0993037*ALOG(TK)+
        1 .0070831558*TK)
          RETURN
          END
          SUBROUTINE PSY1(DB,WB,PB,DP,PV,W,H,V,RH)                   001970
C     THIS ROUTINE CALCULATES' VAPOR PRESSURE PV, HUMIDITY RATIO W, 001980
C         ENTHALPY H, VOLUME V, RELATIVE HUMIDITY RH, AND           001990
C         DEW POINT TEMPERATURE DP(                                 002000
C         WHEN THE DRY BULB TEMPERATURE DB, WET BULB TEMPERATURE WB,002010
C         AND BAROMETRIC PRESSURE PB ARE GIVEN                      002020
C     UNITS' DB, WB, + DP )F]( PB, + PV )IN OF HG]( W)= WATER VAPOR 002030
C         PER = DRY AIR]( H )BTU/= OF DRY AIR]( V )FT**3/= OF DRY   002040
C         AIR( RH IS A FRACTION, NOT (                              002050
          C(F)=(F-32.0E0)/1.8E0                                     002060
```

Figure B.5   (Continued)

```
         PVP=PVSF(WB)                                                    002070
         WSTAR=0.622*PVP/(PB-PVP)                                        002080
         IF (WB.GT.32.0) GO TO 105                                       002090
         PV=PVP-5.704E-4*PB*(DB-WB)/1.8                                  002100
         GO TO 110                                                       002110
   105   CDB=C(DB)                                                       002140
         CWB=C(WB)                                                       002150
         HL=597.31+0.4409*CDB-CWB                                        002160
         CH=0.2402+0.4409*WSTAR                                          002170
         EX=(WSTAR-CH*(CDB-CWB)/HL)/0.622                                002180
         PV=PB*EX/(1.+EX)                                                002190
   110   W=0.622*PV/(PB-PV)                                              002200
         V=0.754*(DB+459.7)*(1.0+7000.0*W/4360.0)/PB                     002210
         H=0.24*DB+(1061.0+0.444*DB)*W                                   002220
         IF (PV.GT.0.0) GO TO 115                                        002230
         PV=0.0                                                          002240
         DP=0.0                                                          002250
         RH=0.0                                                          002260
         RETURN                                                          002270
   115   IF (DB.NE.WB) GO TO 120                                         002280
         DP=DB                                                           002290
         RH=1.0                                                          002300
         RETURN                                                          002310
   120   DP=DPF(PV)                                                      002320
         RH=PV/PVSF(DB)                                                  002330
         RETURN                                                          002340
         END                                                            002350
         FUNCTION PVSF(X)                                                002580
         DIMENSION A(6),B(4),P(4)                                        002590
         DATA A/-7.90298,5.02808,-1.3816E-7,11.344,8.1328E-3,-3.49149/  002600
         DATA B/9.09718,-3.56654,0.876793,0.0060273/                    002610
         T=(X+459.688)/1.8                                              002620
         IF (T.LT.273.16) GO TO 100                                      002630
         Z=373.16/T                                                     002640
         P(1)=A(1)*(Z-1.0)                                               002650
         P(2)=A(2)*ALOG10(Z)                                            002660
         Z1=A(4)*(1.0-1.0/Z)                                            002670
         P(3)=A(3)*(10.0**Z1-1.0)                                       002680
         Z1=A(6)*(Z-1.0)                                                002690
         P(4)=A(5)*(10.0**Z1-1.0)                                       002700
         GO TO 105                                                       002710
   100   Z=273.16/T                                                     002720
         P(1)=B(1)*(Z-1.0)                                               002730
         P(2)=B(2)*ALOG10(Z)                                            002740
         P(3)=B(3)*(1.0-1.0/Z)                                          002750
         P(4)=ALOG10(B(4))                                              002760
   105   SUM=0.0                                                        002770
         DO 110 I=1,4                                                    002780
   110   SUM=SUM+P(I)                                                    002790
         PVSF=29.921*10.0**SUM                                          002800
         RETURN                                                          002810
         END                                                            002820
         FUNCTION DPF(PV)                                                002830
C        THIS ROUTINE CALCULATES DEW-POINT TEMPERATURE FOR A GIVEN       002840
C             VAPOR PRESSURE PV                                          002850
         DP(A,B,C,Y)=A+(B+C*Y)*Y                                         002860
         Y=ALOG(PV)                                                      002870
         IF (PV.GT.0.1836) GO TO 100                                     002880
         DPF=DP(71.98,24.873,0.8927,Y)                                  002890
         RETURN                                                          002900
   100   DPF=DP(79.047,30.579,1.8893,Y)                                 002910
         RETURN                                                          002920
         END                                                            002930
```

Figure B.5   (Continued)

NRC FORM 335
(7-77)

U.S. NUCLEAR REGULATORY COMMISSION
BIBLIOGRAPHIC DATA SHEET

| 1. REPORT NUMBER (Assigned by DDC)
NUREG-0733 |

4. TITLE AND SUBTITLE (Add Volume No., if appropriate)
Analysis of Ultimate-Heat-Sink Spray Ponds

2. (Leave blank)

3. RECIPIENT'S ACCESSION NO.

7. AUTHOR(S)
Richard B. Codell

5. DATE REPORT COMPLETED

MONTH	YEAR
January	1981

9. PERFORMING ORGANIZATION NAME AND MAILING ADDRESS (Include Zip Code)
Hydrologic Engineering Section
Hydrologic Geotechnical Engineering Branch
Division of Engineering, NRR
Washington, D.C. 20555

DATE REPORT ISSUED

MONTH	YEAR
August	1981

6. (Leave blank)

8. (Leave blank)

12. SPONSORING ORGANIZATION NAME AND MAILING ADDRESS (Include Zip Code)

Same as Box 9

10. PROJECT/TASK/WORK UNIT NO.

11. CONTRACT NO.

13. TYPE OF REPORT

Technical Report

PERIOD COVERED (Inclusive dates)

15. SUPPLEMENTARY NOTES
Computer programs available

14. (Leave blank)

16. ABSTRACT (200 words or less)
This report develops models which can be utilized in the design of certain types of spray ponds used in ultimate heat sinks at nuclear power plants, and ways in which the models may be employed to determine the design basis required by U.S. Nuclear Regulatory Commission Regulatory Guide 1.27.

The models of spray-pond performance are based on heat and mass transfer characteristics of drops in an environment whose humidity and velocity have been modified by the presence of the sprays. Drift loss from the sprays is estimated by a ballistics model.

The pond performance model is used first to scan a long-term weather record from a representative meteorological station in order to determine the periods of most adverse meteorology for cooling or evaporation. The identified periods are used in subsequent calculations to actually estimate the design-basis pond temperature. Additionally, methods are presented to correlate limited quantities of onsite data to the longer offsite record, and to estimate the recurrence interval of the design-basis meteorology chosen.

17. KEY WORDS AND DOCUMENT ANALYSIS

17a. DESCRIPTORS

spray ponds
meteorology
ultimate heat sinks

17b. IDENTIFIERS/OPEN-ENDED TERMS
spray ponds, ultimate heat sink (UHS), meteorology, heat transfer

18. AVAILABILITY STATEMENT

19. SECURITY CLASS (This report)
unclassified

21. NO. OF PAGES

20. SECURITY CLASS (This page)
unclassified

22. PRICE
S

NRC FORM 335 (7-77)